Machine Learning Control by Symbolic Regression

Askhat Diveev • Elizaveta Shmalko

Machine Learning Control by Symbolic Regression

 Springer

Askhat Diveev
Federal Research Center
"Computer Science and Control"
Russian Academy of Sciences
(FRC CSC RAS)
Moscow, Russia

Elizaveta Shmalko
Federal Research Center
"Computer Science and Control"
Russian Academy of Sciences
(FRC CSC RAS)
Moscow, Russia

ISBN 978-3-030-83215-5   ISBN 978-3-030-83213-1   (eBook)
https://doi.org/10.1007/978-3-030-83213-1

This Springer imprint is published by the registered company Springer Nature Switzerland AG
The registered company address is: Gewerbestrasse 11, 6330 Cham, Switzerland

Preface

As a reader of this book, you are interested in the creation of modern control systems—automatic, capable of learning, and intelligent—such that you are able to respond to the new challenges of modern technical development, with ubiquitous robotization and digitalization.

The theory of automatic control has gone through a long process of transformation, from a scattered set of control methods for mechanical, hydrodynamic, and other systems to fundamental science. Almost all scientific research in control has been carried out within the framework of studying the possibility of creating various innovative technical solutions from machine tools at the inception stage to automatic flying and space vehicles or nuclear power plants. The twentieth century is famous for the creation of automatic control systems for industrial complexes and production processes using computers. However, the twenty-first century brings new challenges associated with the emergence of universal objects, such as autonomous robots and robotic systems, capable of autonomously performing completely different tasks in different conditions and environments. Modern control systems must be able to quickly change, refine, and learn. This circumstance requires both the universalization and automation of the very process of developing control systems that are not tied to the physics of the control object, operating with laws and patterns that are valid for objects of any complexity and nature. Machine learning meets these new challenges. In this context, a new direction has recently appeared that implements the modern approach of machine learning in the field of control, called Machine Learning Control (MLC).

This book discusses machine approaches to solve control problems.

Today, the most common machine learning apparatus are the various neural networks. At first glance, it seems that a variety of neural network structures can satisfy any control problem. In fact, the structure of the neural network is determined by the developer and it is a given structure in which only parameters are configured, while the structure itself remains unchanged. So it is difficult to even guess whether this structure is optimal for a given task. In addition, for complex tasks, a neural network has a complex structure, and for a development engineer who is used to describing objects and systems with some functions that have physical meaning and geometric representation, working with a neural network seems to be a kind of black box.

This book aims to introduce to a wide range of readers to the possibilities of symbolic regression methods. Symbolic regression methods allow us to find functions in a form that engineers are familiar with. These methods can be considered as the universal tool for solving machine learning control problems. The variety of symbolic regression methods can automate the process of synthesis of control systems, but very few of them are used in this direction. This is due to a number of difficulties, such as non-numerical search space and the absence of a metric on it, the complexity of the program code and the absence of publicly available software packages, etc. Thus, the main purpose of this book is to show the accumulated world experience in the field of symbolic regression methods and their possibilities in terms of application to control theory and practice.

The more widely we distribute these technologies, the more valuable they will become. To unleash the potential of symbolic regression methods, we want to make them available to a wide range of researchers and applied engineers. Machine Learning Control by Symbolic Regression is written in a simple language accessible to a wide range of readers, but it does not ignore the complex points and mathematically rigorous formulations and justifications. Control system engineers, mathematicians, specialists in optimization, specialists in mathematics and combinatorics, machine learning specialists, software developers, and graduate students will find a lot of value in the pages of this book.

Moscow, Russia   Askhat Diveev
April, 2021   Elizaveta Shmalko

Contents

Acronyms

ML	Machine learning
MLC	Machine learning control
ANN	Artificial neural network
NN	Neural network
ADOC	Analytical design of optimal controllers
ADAR	Analytical design of aggregated regulators
GA	Genetic algorithm
GP	Genetic programming
GE	Grammatical evolution
CGP	Cartesian genetic programming
IGP	Inductive genetic programming
NOP	Network operator
AP	Analytic programming
PME	Parse-matrix evolution
BCGP	Binary complete genetic programming
MNOP	Multilayer network operator

Chapter 1
Introduction

Abstract This book is primarily about control. In the introduction, we will talk about modern approaches to control, about the automation of the very design process of control, about artificial intelligence and machine learning, and, of course, about symbolic regression methods, which open up new possibilities not only in the field of control automation, but also in the design of completely different optimal structures, including building structures, technical systems, and even musical works.

1.1 About Modern Control Systems

The process of making human labor easier by technical means is entering a new phase. Until recently, the emergence of new means of automation was mainly limited to the elimination of physical labor from production processes, which made it possible to increase accuracy, safety, economy, and productivity. In many industries where the process was sufficiently stable, manual work was replaced by fully automated lines. Today, outstanding breakthroughs in computer technology, as well as fantastic advances in robotic technology, trigger new changes not only in the field of automation, but also in the very process of designing control systems for automatic devices.

At the present stage, since the 1980s, a computer plays the role of a control device. Thus, digital automation, or digital automatic control, can be defined as a technology that uses programmed commands affecting some object or process, and feedback, with the help of which it is determined whether these commands are executed correctly [1]. And the development of the control system now consists in programming the control device.

With the advent of high-performance and relatively inexpensive computers, the very approaches to the development of control systems have changed. Previously, the development of a control system consisted in the initial formation of the system configuration (setting a certain structure) and further calculation and adjustment of

A. Diveev, E. Shmalko, *Machine Learning Control by Symbolic Regression*,
https://doi.org/10.1007/978-3-030-83213-1_1

optimal parameters that would provide the desired quality indicators. The stage of setting the configuration of the control system for a particular control object was akin to art, in the process of which the experience and intuition of the development engineer has a colossal role. However, the demands of the times have changed.

Modern control theory deals with automation and robotization that must meet the requirements of adaptability, robustness, and optimality. The variability of modern control objects creates the need to develop universal approaches to the synthesis of control systems; the very process of developing control systems is required to be automated. The number and variety of robots is growing at an enormous rate. So are we going to program each robot by hand?

The question arises: Can we trust the machine on its own, without the help of an engineer specifying the structure of the control system, to automatically determine this structure by examining the data or relying on given estimates or desired characteristics of the system? This question opens the door to a new paradigm for the development of control systems: machine learning control [2].

1.2 About Machine Learning Control

Machine learning is one of the areas of artificial intelligence associated with solving problems based on algorithms that can learn or gradually improve the performance of a given task. Machine learning is based on the idea that computing systems are able to show a behavior that was not explicitly programmed in them, they can identify patterns, rules, or functional dependencies and make decisions on their own. Control systems can act as such functional dependencies. In machine learning, the control system is learned, not programmed by the developer.

The most famous machine learning technique now is neural networks, and sometimes machine learning is equated with neural network training. This is incorrect, because machine learning is a broader concept [3–5]. It includes such early forms of data analysis as probabilistic modeling based on the application of Bayes' theorem and logistic regression; classification algorithms such as kernel methods, hierarchical structures such as decision trees, random forests, and gradient boosting. We definitely note that deep neural networks show the best performance in many tasks, which explains their popularity. At its core, a neural network is a function with a specific structure and a large number of unknown parameters. Learning, or, more precisely, training neural networks is finding optimal values of parameters. But despite the successful breakthroughs in the development of artificial neural networks, they have their drawbacks [6, 7]. First, the choice of the type and structure of the neural network, determination of the size and number of layers, and so on are carried out mostly intuitively, at the discretion of the developer, based on his experience and knowledge. Another big drawback of neural networks is the lack of interpretability of the received functional dependency, since in fact the received function in this approach is a black box, which gives little opportunity for its understanding and analysis. Moreover, approaches based on deep neural networks often suffer from lack of reproducibility, caused largely by non-determinism in the learning process. Finally, an important limitation of neural networks is the large amount of data required to train them. In many problems, including the problem of optimal control synthesis, such a volume of training data simply does not exist.

With this in mind, more and more works appear on the study of other artificial intelligence methods for machine learning. One of these areas is symbolic regression methods.

1.3 About Symbolic Regression Methods

Symbolic regression is a relatively new mathematical or algorithmic construction that emerged at the end of the twentieth century and was originally intended to solve the programming problem of automatically writing software code. In this task, one program searches by some criterion and generates the code of another program to solve a given problem. The required program code is written in a universal form from a set of prefix operators, the code of which contains the operator identifier and operand identifiers, among which there may be identifiers of other operators. Such a construction is schematically a tree, the nodes of which are associated with operators, and the number of branches leaving each node is equal to the number of operands of this operator. The first symbolic regression method, a genetic programming, authored by John Koza [8] has such construction. The merit of John Koza, who was one of doctoral students of the famous John Holland, the creator of the genetic algorithm, lies in the fact that he was able to use the genetic algorithm to find the optimal code with a structure in the form of a tree. For this purpose, it was necessary to redefine the main operation of the genetic algorithm, the operation of crossover. In the genetic programming, in contrast to the genetic algorithm, the crossover of two codes in the form of a tree is performed using the exchange of subtrees.

Note right away that this book does not consider the problem of automatic writing of programs. The basic brilliant idea of genetic programming is to apply a genetic algorithm to find a solution in the form of a code. Genetic algorithm, one of the few optimization algorithms that does not use arithmetic operations, addition and multiplication in the search process. These operations are not applicable to different codes.

Anything can be encoded, from a piece of music to the design of a spacecraft. Humanity has been coding its activities for a long time, possibly with the aim of transferring experience to future generations. In this case, the codes of the desired structure must be written in a form convenient for processing on a computer. And then it is necessary to develop the crossover operation for codes so that as a result of its implementation, new correct codes corresponding to new constructions are obtained. It should be stipulated here that when searching for an optimal solution, it is also necessary to determine a numerical estimate for each possible solution. In some problems, this is absolutely not difficult to do, for example, if you look for a code for a mathematical expression for solving some algebraic or differential equation. In other tasks, it is almost impossible to do this, for example, if you search for the code of a piece of music. In this case, the problem of coding music is not particularly difficult, since people have long learned to encode sounds, and in a form that can be easily represented on a computer. Here the main problem lies precisely in the assessment of the desired solution, which may be liked by some people and not by others.

In this book, symbolic regression is used to find mathematical expressions of functional dependencies. Problems from the field of control theory are considered. An attempt to solve classical control problems by new methods is presented. Therefore, this book is primarily intended for specialists in the field of control. After reading several chapters of this book, we would like control specialists, theorists, and engineers, to rush to computers willing to program the application of symbolic regression methods to solve complex control problems that have long tormented them. It also aims to be a useful reference guide for applying symbolic regression techniques to control problems.

It should be noted that in addition to genetic programming, there are many other symbolic regression methods that differ in the form of coding and, accordingly, in the crossover operation of the genetic algorithm. We tried to include in the book descriptions of all the currently known methods of symbolic regression in order to be able to freely use them in solving applied problems of synthesis of control of various objects.

We expect that other readers of the book should be experts in the field of machine learning. The book formulates the general formal statement of the machine learning problem as the problem of finding an unknown function. In all areas where such a problem arises, it is possible to use machine learning methods. Previously, in the overwhelming majority of cases, to search for functions, a researcher intuitively wrote it down with an accuracy to parameters. Then, as a rule, the values of these parameters were found by the least squares method. Note that neural networks, which are now the main tool of machine learning, are also functions that are defined accurate to the values of a large number of parameters. Even in those cases when the structure of a neural network is being looked for, the change in its structure is also determined by the values of regular parameters, the number of layers, or the number of neurons in a layer. Symbolic regression methods, in contrast to neural networks, allow finding not only parameters, but also the structure of the mathematical expression of a function.

In the field of control, the problem of finding a function is encountered in the problems of optimal control, or generally in control synthesis, and identification of the mathematical model of the control object. The formulation of these problems and known methods for their solution are described in Chap. 2.

In the pioneer book on machine learning control [2], the control synthesis problem is solved by the genetic programming method. However, the example of control synthesis given in the book consists in finding the controller parameters by a genetic algorithm, i.e. again the search for the values of the parameters of a function with a given structure. In this book, a general description of the symbolic regression methods is presented in Chap. 3, and then not only the genetic programming but also various other symbolic regression methods are described in detail (Chap. 4), and examples of solving control synthesis problems by these methods are given in Chap. 5.

Note that almost twenty years of experience in using symbolic regression methods to solve the control problems led us to the definition of the main problem of these methods, which does not yet allow them to effectively find formal analytical solutions to various mathematical problems. This is the problem of finding an optimal solution in a code space that does not have a single numerical measure. As in the space of words: there is an alphabet and words can be close, based on the assessment of symbols, but have completely different meanings, based on the semantic assessment. The proximity of the names does not correspond to the proximity of the meanings. The same is the case with the search on the space of function codes. Function evaluation works with mappings. And the search is carried out on codes, that is, on the names of these mappings. Thus, the metric between the names of the functions does not correspond to the distances between the values of the functions.

In general, the encoding of a mathematical expression by any symbolic regression method can be described as follows. The basic set of elementary functions and their code representation are determined. This basic set is like an alphabet. It can also include the arguments of the desired mathematical expression. Further, the rules for forming words (multisets of codes) from this alphabet are determined. Each word is a code of the desired mathematical expression, which is inserted into the mathematical problem to be solved and evaluated using a certain functional. The problem is that the search for optimal solution is organized on the non-numerical space of codes where only some symbolic metric can be set such as the Levenshtein, Hamming, or Jaro distance, but the estimation of the solutions during the search is performed in the space of functions with absolutely another metric. It turns out that the search process is carried out on the space of function codes, where there is no single metric, and it is impossible to determine a numerical estimate of the distance between any two different possible solutions. It is obvious that optimization algorithms, which use arithmetic operations, cannot be used to find a solution here. Among known optimization algorithms, only the genetic algorithm [9, 10] does not use arithmetic operations and allows working with codes. However, it is not always effective due to the complexity of the search space and requires additional improvements. We managed to improve the efficiency of the genetic algorithm by applying the principle of small variations in the basic solution, which in the general case defines some conditions for the form of coding. A detailed discussion of the search problems in non-numerical space of codes, the application of genetic algorithm, and the concept of the principle of small variations are presented in Chap. 3. We do not exclude the possibility of using some other approaches here. The need to develop new modern algorithms for finding solutions is obvious; breakthrough results can be obtained here, therefore the attention of a wide range of researchers is very important. If we draw an analogy with the technologies of neural networks, then their structure has not changed much since their inception. But the main breakthroughs were achieved with the advent of new learning methods, primarily the backpropagation. We hope that the readers of this book will also be mathematicians dealing with the problems of creating efficient algorithms for solving NP-hard problems. They will be able to find here a new class of complex optimization problems, which is defined as the problem of finding the optimal code. And new, more "reasonable," intelligent search optimization algorithms will be developed.

The material of the book is presented in a form accessible to a wide range of readers, combining both classical mathematical formulations and substantiations of the problems under consideration, as well as explanations and algorithms that are understandable for programming.

References

1. Dorf, R.C., Bishop R.H.: Modern Control Systems, 12th edn. Prentice Hall, Upper Saddle River (2011)
2. Duriez, T., Brunton, S.L., Noack B.R.: Machine Learning Control–Taming Nonlinear Dynamics and Turbulence. Springer International Publishing, Switzerland (2017)
3. Bishop, C.M.: Pattern Recognition and Machine Learning. Springer, New York (2006)
4. Murphy, K.P.: Machine Learning: A Probabilistic Perspective. MIT Press, Cambridge, MA (2012)
5. Goodfellow, I., Bengio, Y., Courville A.: Deep Learning. MIT Press, Cambridge/London (2016)
6. Jiayuan, M., Chuang, G., Pushmeet, K., Joshua, T., Jiajun, W.: The neuro-symbolic concept learner: interpreting scenes, words, and sentences from natural supervision. In: International Conference on Learning Representations (2019)
7. Nagarajan, P., Warnell, G., Stone., P.: The impact of nondeterminism on reproducibility in deep reinforcement learning (2018)
8. Koza, J.R.: Genetic Programming: On the Programming of Computers by Means of Natural Selection. MIT Press, Cambridge, MA/London (1992)
9. Goldberg, D.: Genetic Algorithms in Search, Optimization and Machine Learning. Addison-Wesley Professional, Reading (1989)
10. Mitchell, M.: An Introduction to Genetic Algorithms. MIT Press, Cambridge, MA (1996)

Chapter 2
Mathematical Statements of MLC Problems

Abstract This chapter presents the formal statements of MLC problems. First of all, consider the formulation of the machine learning problem as the problem of finding an unknown functional relationship. Next, we present the formulations of control theory problems that can be distinguished as machine learning control problems, namely the optimal control problem and more widely the general control synthesis problem, optimal control problem based on the synthesis of the stabilization system (synthesized optimal control), and the control object identification problem. All the tasks involve finding an unknown function. The function can be set up to parameters, and then machine learning techniques are used only to adjust the parameters. In general case, both the structure of the function and its parameters should be found.

2.1 Machine Learning Problem

In many, if not all, scientific disciplines, the main task of research is to find a functional relationship between certain values of parameters that characterize the properties of the object under research. If it is possible to present the sought functional dependence in the form of a mathematical formula, then very often such a formula becomes a law in this area and acquires the name of its creator. Here we would like to emphasize the importance of the process of finding a mathematical expression for a function.

Machine learning, in almost all known cases, is the search for some functional dependence between the values of certain quantities. Unlike symbolic regression methods, which is the main focus of the present book, machine learning by neural networks, for example, also allows finding a functional relationship between some characteristics, but in the form of a computational black box. The resulting functional dependence can be used for modeling, prediction, classification, etc., but it is impossible to determine the mathematical expression of this function. Hence, the significance of the result obtained is somewhat different. It can be argued that the

A. Diveev, E. Shmalko, *Machine Learning Control by Symbolic Regression*,
https://doi.org/10.1007/978-3-030-83213-1_2

mathematical formula for some functional dependence is the crown of research and establishes the fact that the researcher has reached a certain level of knowledge in understanding the process and now he can share this knowledge in the form of a formula with humanity.

To formulate mathematically a machine learning problem, it is necessary to assume that in the process under study there is a functional relationship between the values of some parameters of this process. This functional relationship allows to determine the values of some parameters, called output, basing on the values of other parameters, called input. At the same time, the mathematical formula describing the implementation of this functional relationship is unknown. The values of the input and output vectors can be determined as a result of experiments.

Definition 2.1. A set of computational procedures, which transforms a vector $\mathbf{x}$ from an input space X into a vector $\mathbf{y}$ from an output space Y, and in which there is not any mathematical expression $y = f(x)$ for them, is called an unknown function.

Denote the unknown function between input vector $\mathbf{x}$ and output vector $\mathbf{y}$ as

$$\mathbf{y} = \alpha(\mathbf{x}). \tag{2.1}$$

Definition 2.2. Machine learning is the process of computer implementation of a computational procedure for finding an unknown function.

We believe that such a definition of machine learning fully covers the main tasks that are currently being solved using machine learning: approximation, forecasting, clustering, etc. For example, a classification problem can also be thought of as finding a function that has an integer step-wise character.

Now, in order to apply machine learning to solve different problems, it is necessary to formulate these problems as problems of finding an unknown function

$$\mathbf{y} = \beta(\mathbf{x}, \mathbf{q}), \tag{2.2}$$

where $\mathbf{q}$ is the vector of the required parameters, $\mathbf{q} \in \mathbb{R}^p$, and β is some function that is equal or close to α in terms of some criterion.

There are two types of approaches of searching for an unknown function: parametric and structural-parametric.

Type 1   The first approach is parametric, when the structure of the unknown function is determined by the researcher accurate to the values of a certain number of parameters, i.e. in (2.2) β is given.

Machine search for an unknown function in this case consists of finding the optimal parameter values $\mathbf{q}$ according to a given criterion.

The parametric approach includes also the case when the structure of a function regularly changes during the search, for example, if the function is searched in the form of a mathematical series and the number of members of the series is determined during the search.

Searching for an unknown function basing on artificial neural networks also refers to the parametric approach. Indeed, transformations performed in any neural network are formally described by a function with a given structure and a large number of unknown parameters.

Type 2   Another approach is `structural-parametric`. According to this approach, the machine search for an unknown function consists of finding the optimal structure of the function β and the optimal value of the vector of parameters $\mathbf{q}$.

Today, the structural-parametric approach can be implemented using symbolic regression methods. These methods define a basic set of primitive functions and coding rules. Then, using a genetic algorithm, the optimal structure of the mathematical expression of the desired function on the code space together with optimal parameters is searched. Symbolic regression methods differ in coding rules and crossover and mutation operations of the genetic algorithm performed on codes.

Further in this book we take a detailed tour of the different approaches to machine learning based on the application of symbolic regression methods. Ultimately the idea is to create a clear map of the vast landscape of symbolic regression methods.

The search for an unknown function, which, as we have already defined, is the goal of machine learning, should be carried out on the basis of some evaluating criterion. Depending on the type of evaluating criterion, machine learning problems can be divided into two main classes: unsupervised learning and supervised learning. Note that the currently existing various types of machine learning can be attributed to one of these categories according to evaluating criterion.

In some problems, an evaluating criterion is specified

$$\gamma(\beta(\mathbf{x}, \mathbf{q})) : X \times \mathbb{R}^p \to \mathbb{R}^1. \tag{2.3}$$

Particularly, such evaluating criterion for machine learning control problems can be a quality functional. For example, in the control synthesis problem considered in more detail below it is necessary to find the control function dependent of the coordinates of the state space of the object. To estimate the desired function, it is substituted into the model of the control object and the value of the given functional is calculated. The researcher usually knows the approximate value of the functional for the optimal solution of the problem and set it up to evaluate the received solution.

Definition 2.3. `Unsupervised machine learning` consists of finding a function (2.2) such that for some given estimate (2.3) the following inequation is true

$$\|f^* - \gamma(\beta(\mathbf{x}, \mathbf{q}))\| \le \delta, \tag{2.4}$$

where f^* is a satisfactory value of the estimate, and δ is a small positive value.

Another approach for evaluating the desired function is to create a training set. A training set is some possible examples that are used during the learning process to search for unknown function.

Definition 2.4. A pair of sets of compatible dimensions

$$(\tilde{X}, \tilde{Y}) \tag{2.5}$$

is called a `training set` if

$$\tilde{X} = \{\mathbf{x}^1, \ldots, \mathbf{x}^N\} \subseteq X, \tag{2.6}$$

$$\tilde{Y} = \{\mathbf{y}^1 = \alpha(\mathbf{x}^1), \ldots, \mathbf{y}^N = \alpha(\mathbf{x}^N)\} \subseteq Y, \tag{2.7}$$

and it is assumed that there is one-to-one mapping $X \to Y$.

Definition 2.5. `Supervised machine learning` consists of building a training set (2.5) and finding a function (2.2) such that if the total error for the training set is less than the given value ε

$$\sum_{i=1}^{N} \|\mathbf{y}^i - \beta(\mathbf{x}^i, \mathbf{q})\| \leq \varepsilon, \tag{2.8}$$

then for $\forall \mathbf{x}^*$ not included in the training set $\mathbf{x}^* \notin \tilde{X}$ the following inequation is fulfilled

$$\|\mathbf{y}^* - \beta(\mathbf{x}^*, \mathbf{q})\| \leq \delta, \tag{2.9}$$

where $\mathbf{y}^* = \alpha(\mathbf{x}^*)$.

Based on the introduced formulations and strict mathematical definitions, we can now apply machine learning methods in various tasks where it is required to search for a function. A vast area of such problems are control tasks.

Definition 2.6. `Machine learning control` is a machine search for an unknown control function using machine learning methods.

Such problems in the field of control include the problem of optimal control in various formulations, for example, in the formulation of Pontryagin or Bellman, the problem of general synthesis of control, as a feedback function on the state of the object, as well as the problem of identifying the model of the control object itself, as a search for functions of the right-hand sides of the equations of dynamics of the object. All these tasks require finding an unknown function, and therefore, machine learning methods can be applied to them.

In the mathematical formulation of control problems, the model of a control object is described by a system of ordinary differential equations. The right-hand sides of this system of differential equations, written in the Cauchy form, include a free control vector. This vector is called "free" because it can take any values from a certain limited set of values. Without specific values of the control vector, the system of ordinary differential equations describing the mathematical model of the control object cannot be solved. Object control in the classical mathematical sense is to qualitatively change the right-hand sides of the differential equations due to the control vector included in them. Such description is used in the next sections in formulation of control problems as MLC problems.

2.2 Optimal Control Problem

The optimal control problem is the most famous problem in the field of control theory. This problem at one time attracted mathematicians to the field of control and made control theory a branch of mathematics.

Previously in the field of control, applied methods based on the use of transfer functions or frequency characteristics, which were borrowed from electrical engineering, prevailed. According to the frequency approach, each element of the control system was considered as a low-pass filter. The developer needed to build such a control system from separate blocks with certain frequency characteristics, so that, as a result, the control system, together with the object, would pass useful signals without distortion and, if possible, would not allow interference.

The representation of the mathematical model of the control object in the form of a system of ordinary differential equations began to be widely used after the formulation of the optimal control problem. To emphasize the adherence to modern trends, and not to the outdated apparatus of transfer functions, in the sixties of the last century many authors wrote in scientific articles that the state space method was used when considering the control system. At the present time, the problem of optimal control continues to be actively studied by mathematicians all over the world, probably because an effective general numerical method has not yet been developed for its solution.

In the optimal control problem, the control object is described by a system of ordinary differential equations, in the right-hand sides of which there is an unknown control vector. The initial and terminal conditions and the integral quality functional are also given. It is necessary to find control as a function of time. If we substitute this function into the right-hand sides of the differential equations, then we obtain a non-stationary system of differential equations with a known function of time on the right-hand side. A particular solution of this non-stationary system of differential equations from the given initial conditions reaches terminal conditions, and the value of the quality functional is optimal in this case.

The optimal control problem belongs to the problems of infinite-dimensional optimization, since it is necessary to find a vector function of time, and not a constant vector in a real vector space of a certain dimension. Therefore, this problem must belong to the class of problems of the calculus of variations. Initially, attempts were made to solve the optimal control problem using the calculus of variations. However, it turned out that the constraints on the values of control and the possibility of the control function to have discontinuities of the first kind both in value and in derivatives make it impossible to apply the methods of the calculus of variations to the optimal control problem.

At present, the most significant result in the field of optimal control is the Pontryagin maximum principle. Application of this approach to the optimal control problem allows to consider the optimal control problem as an optimization problem in a finite-dimensional space. According to the maximum principle, conjugate variables are introduced into the problem, which in this case play the role of Lagrange multipliers in the conditional minimization problem. Conjugate variables are necessary in order to take into account the connections described by the mathematical model of the control object when minimizing a given functional. Unlike Lagrange multipliers, conjugate variables change over time, and to determine them, it is necessary to solve a system of differential equations. Next, the Hamiltonian is constructed, which includes the functional, conjugate variables and the right-hand sides of the system of differential equations of the control object model. According to the maximum principle, a necessary condition for optimal control is to maximize the value of the Hamiltonian at each time instant. In order to find the optimal control from the condition of the maximum of the Hamiltonian, it is necessary at each moment of time to know the value of the state vector of the control object and the vector of conjugate variables. The problem is that the initial and terminal conditions are known for the system of differential equations of the plant model, and for conjugate variables there is a system of differential equations, but this system has neither initial nor terminal conditions. Therefore, a boundary value problem arises of finding the initial conditions for a system of conjugate variables such that, when calculating the control at each integration step, from the conditions for the maximum of the Hamiltonian, the state vector as a result falls into the given terminal conditions. This problem is formulated as a finite-dimensional optimization problem, in which it is necessary to find the vector of initial conditions for a system of differential equations of conjugate variables according to the criterion of the minimum error of the state vector entering the terminal conditions.

There are also direct numerical methods that can be used for the optimal control problem, without applying the Pontryagin maximum principle, directly according to a given integral quality criterion.

The optimal control problem is presented here for several reasons. First, in the optimal control problem, it is necessary to find a function, albeit one variable, but this means that when searching for a function, machine learning methods can be used. Second, after solving the optimal control problem and finding the control as a function of time for the practical implementation of the found optimal solution, it is necessary to construct a system for stabilizing the motion of the control object along the found optimal program trajectory, which leads to the problem of finding another control function and, therefore, again to the problem of machine learning. And finally, the problem of optimal control can be solved after solving the problem of stabilizing the object with respect to the equilibrium point in the state space. This method will be described below, and it leads to finding a solution to the optimal control problem in the class of practically realizable control functions.

Consider a mathematical problem statement of the optimal control problem.

A model of the control object is given in the form of the system of ordinary differential equations

$$\dot{\mathbf{x}} = \mathbf{f}(\mathbf{x}, \mathbf{u}), \tag{2.10}$$

where $\mathbf{x}$ is a vector of the state space, $\mathbf{x} \in \mathbb{R}^n$, $\mathbf{u}$ is a vector of control, $\mathbf{u} \in U \subseteq \mathbb{R}^m$, and U is a compact set, $m \leq n$.

For the system (2.10) initial conditions are given by

$$\mathbf{x}(0) = \mathbf{x}^0. \tag{2.11}$$

Terminal conditions are given by

$$\mathbf{x}(t_f) = \mathbf{x}^f, \tag{2.12}$$

where t_f is a terminal time, which is not given but is determined by achieving the terminal conditions.

The quality criterion is given in the form of an integral and/or terminal functional

$$J = F(\mathbf{x}(t_f)) + \int_0^{t_f} f_0(\mathbf{x}(t), \mathbf{u}(t)) dt \to \min. \tag{2.13}$$

It is necessary to find a control as a function of time

$$\mathbf{u} = \mathbf{v}(t), \tag{2.14}$$

where $\mathbf{v}(t) \in U$ for $t \in [0; t_f]$.

The received control function $\mathbf{v}(t)$ is called a program control. If this program control (2.14) is substituted into the right part of the system (2.10), then the following system of differential equations is obtained

$$\dot{\mathbf{x}} = \mathbf{f}(\mathbf{x}, \mathbf{v}(t)). \tag{2.15}$$

Such system (2.31) has a partial solution $\mathbf{x}(t, \mathbf{x}^0)$ from the initial conditions (2.11), which achieves the terminal conditions (2.12) with an optimal value of the quality criterion (2.30).

Computational methods for the optimal control problem include two approaches.

The first one is called a `direct approach`. It includes methods that search for optimal control in the form of function of time

$$\mathbf{u} = \mathbf{v}(t, \mathbf{q}), \tag{2.16}$$

where $\mathbf{q}$ is a vector of parameters, $\mathbf{q} \in \mathbb{R}^p$.

The optimal control problem here is reduced to a nonlinear programming problem [1, 2] that provides the transition from the optimization problem in the infinite-dimensional space to the optimization problem in the finite-dimensional space. As a result we get a highly dimensional nonlinear programming problem that could be solved using modern methods for solving the problem of nonlinear programming, for example, stochastic gradient search methods, which are successfully used today for training neural networks, but in this case we must be sure that the objective function is unimodal. Unfortunately, most applied optimal control problems have a non-unimodal functional, especially problems with phase constraints, which is often encountered robots control, where each robot is a phase constraint for other robots. This circumstance determines the applicability of both evolutionary algorithms, from the point of view of parametric search for a control function, and symbolic regression methods, as a tool for structural-parametric search.

The second approach is called `indirect` and includes application of the Pontryagin's maximum principle [3]. According to the approach a Hamiltonian is written for the problem

$$H(\mathbf{x},\ \psi,\mathbf{u}) = -\frac{\partial F(\mathbf{x})}{\partial \mathbf{x}}\mathbf{f}(\mathbf{x},\mathbf{u}) - f_0(\mathbf{x},\mathbf{u}) + \psi^T\mathbf{f}(\mathbf{x},\mathbf{u}), \qquad (2.17)$$

where ψ is a vector of conjugate variables, $\psi = [\psi_1 \ \ldots \ \psi_n]^T$.

The vector of conjugate variables is determined from the system of differential equations

$$\dot{\psi} = -\frac{\partial H(\mathbf{x},\ \psi,\mathbf{u})}{\partial \mathbf{x}}. \qquad (2.18)$$

According to Pontryagin's maximum principle an optimal control in each moment provides maximum of Hamiltonian

$$\mathbf{u}(t) = \arg\max_{\mathbf{u}\in U} H(\mathbf{x}(t,\mathbf{x}^0),\ \psi(t),\mathbf{u}), \qquad (2.19)$$

where $\psi(t)$ is a solution of the system (2.18).

The Pontryagin's maximum principle transforms the optimal control problem, where it is necessary to find control function, into the boundary problem, where it is necessary to find a vector of initial conditions for conjugate variables. Application of the maximum principle doubles the dimension of the system due to the introduction of conjugate variables, which requires additional computational costs. Additional difficulty is that for a system of equations (2.18) for conjugate variables, initial conditions are unknown. The approach itself was developed in the 1960s, when computer technology was very different from the modern one and it was important to be able to build analytical solutions, at least for problems of small dimension. Current trends are such that computer technology and numerical approaches are gradually crowding out analytic.

Thus, in the presented mathematical formulation of the optimal control problem, it is required to find the optimal control function (2.14). This means that this problem can be considered as a machine learning control problem and can be solved by machine learning methods.

2.3 Control Synthesis Problem

The control synthesis problem is the main one in the control theory. In contrast to the aboveconsidered optimal control problem, it has a more applied character, since control is sought here as a function of the object state. As a result, the developer receives a feedback control unit, which ensures, according to signals from sensors that determine the object's state, that the object achieves the control goal with the optimal value of the control quality criterion for any current state of the object. This is the specific feature of the control synthesis problem. Solving one control synthesis problem is equivalent to solving an infinite set of optimal control problems. After solving the control synthesis problem, the resulting control unit automatically solves the optimal control problem for any current state of the control object.

At the dawn of control theory creation in the 1960s of the last century, while studying the mathematical formulation of the optimal control problem, R. Bellman formulated the control synthesis problem and derived the Bellman equation [4]. The equation is a partial differential equation. The solution to this equation is the Bellman function, one of the arguments of which is the control vector. Finding such control that maximizes the Bellman function is a solution to the synthesis problem. Note that partial differential equations are much more complicated than ordinary differential equations and in the general case almost never have a common solution. Bellman proposed a numerical procedure for finding a solution in the form of dynamic programming [5,6]. As a result of applying this procedure for a huge number of numerical values of state vectors, we obtain a huge number of control vectors, while we do not obtain any analytical dependence of control on the state.

Other attempts to solve the Bellman equation consider special cases and for them obtain an analytical formula for the Bellman function. Such cases, for example, include linear control systems with a quadratic quality functional. In this case, a control is searched in the form of a linear dependence on the state space coordinates. Such approach of analytical design of optimal controllers (ADOC) [7] is well formalized, but it works only for a narrow class of problems.

At that very time, several control synthesis problems were completely solved based on the Pontryagin maximum principle [3]. It turned out well since simple models of control objects were considered, mainly of the second order. The time-optimal problem was solved. And it managed to obtain general solutions for the differential equations of the control object and conjugate variables. Then, on the basis of the constructed solutions from different initial conditions, the control switching points were determined. As seen, this approach is not universal, but when applying this approach Boltyanskii [8] formulated the problem of general synthesis of control, which is an urgent mathematical problem up to the present time because its mathematical formulation has no general analytical or numerical methods of solution till now.

Consider a conventional formulation of the control synthesis problem.

Given a control object in the form (2.10) of the system of differential equations. The domain of initial conditions in the state space is given by

$$X_0 \subseteq \mathbb{R}^n. \tag{2.20}$$

The existence of the initial condition domain is a main feature of the control synthesis problem. Initially, Boltyanskii defined the domain of initial conditions as a whole space of states $X_0 = \mathbb{R}^n$ and tried to solve this problem analytically. As from practical and computational sense, the initial domain should still be limited. Therefore, we consider the domain X_0 as a restricted set in the space of states.

The terminal conditions (2.12) are given.

The quality criterion is given by

$$J_1 = \int \cdots \int_{X_0} \left(F(\mathbf{x}(t_f, \mathbf{x}^0)) + \int_0^{t_f} f_0(\mathbf{x}(t, \mathbf{x}^0), \mathbf{u}(t)) dt \right) dx_1^0 \ldots dx_n^0 \to \min_{\mathbf{u} \in U}, \quad (2.21)$$

where $\mathbf{x}^0 = [x_1^0 \ldots x_n^0]^T \in X_0$, t_f is not given but determined by achieving the terminal conditions (2.12), it can be different for different initial conditions.

It is necessary to find a control function as a function of the state space vector

$$\mathbf{u} = \mathbf{h}(\mathbf{x}) \in U, \quad \mathbf{h}(\mathbf{x}) : \mathbb{R}^n \to \mathbb{R}^m. \quad (2.22)$$

If the obtained control function is inserted into the right part of the mathematical model (2.10), then the obtained system of differential equations

$$\dot{\mathbf{x}} = \mathbf{f}(\mathbf{x}, \mathbf{h}(\mathbf{x})) \quad (2.23)$$

will have a partial solution for any initial condition from the initial domain (2.20)

$$\mathbf{x}(0) = \mathbf{x}^0 \in X_0, \quad (2.24)$$

which achieves the terminal condition (2.12) with an optimal value of the quality criterion (2.30).

Thus, solving the synthesis problem as finding the control function (2.22) corresponds to the machine learning control.

For computational solution of the control synthesis problem (2.10), (2.20), (2.12), (2.21), the initial condition domain (2.20) is replaced by a finite set of initial conditions

$$X = \{\mathbf{x}^{0,1}, \ldots, \mathbf{x}^{0,K}\}, \quad (2.25)$$

and the multiple integral of the quality criterion (2.21) is replaced by corresponding sum for all initial conditions

$$J_2 = \sum_{i=1}^{K} \left(F(\mathbf{x}(t_{f,i}, \mathbf{x}^{0,i})) + \int_0^{t_{f,i}} f_0(\mathbf{x}(t, \mathbf{x}^{0,i}), \mathbf{u}(t)) dt \right), \quad (2.26)$$

where $t_{f,i}$ is the time of achieving the terminal condition from the initial condition $\mathbf{x}^{0,i}$, $i = 1, \ldots, K$.

In the search process the time of achieving the terminal condition is determined by the following equation:

$$t_{f,i} = \begin{cases} t, & \text{if } t < t^+ \text{ and } \|\mathbf{x}^f - \mathbf{x}\| \le \varepsilon \\ t^+, & \text{otherwise} \end{cases}, \quad (2.27)$$

where ε is an accuracy of achieving the terminal condition, t^+ is a limit time for achieving the terminal condition, and ε and t^+ are given positive numbers.

The problem of the numerical solution of the control synthesis problem is to find the structure of the mathematical expression of the multidimensional control function and its parameters. The dimension of the control function is generally equal to the dimension of the control vector, and the number of arguments of the control function is equal to the dimension of the state vector.

Since the synthesis problem is extremely important in control theory, a large baggage of methods for its solution has already been accumulated. Let us take a quick look at the existing approaches.

The most famous analytical method for solving the control synthesis problem is the integrator backstepping, developed in 1992 by Kokotovic [9, 10]. The essence of this method is to include some nonlinearities into the control function based on the analysis of the right-hand sides of the differential equations in order to compensate them and obtain a constant sign Lyapunov function for a closed control system, for example, with even powers of the state vector components, and with the same sign. The method is implemented manually by the researcher, depends on the model of the control object, and works especially well for cascade systems in which some coordinates of the state vector are control for other coordinates, for example, in some aircrafts the spatial movement is controlled by the angular position. The application of this method is rather effective for low-order systems.

Similar in concept is the method of analytical design of aggregated regulators (ADAR), developed by Kolesnikov [11, 12]. The method consists of introducing aggregated variables that describe the control goal, for example, the terminal state. These variables are introduced into the functional and then, when composing the Bellman equation, the time derivative is taken with respect to them. In the analytical calculation of derivatives, the aggregated variables include the right parts of the control object model; thus, they depend on the control vector. Then we get a system of nonlinear equations, the number of which is almost always equal to the dimension of the state vector. These equations include the control vector. By solving these equations with respect to the control vector, we obtain the control function as a function of the coordinates of the state space. There are tasks in which the method turns out to be effective; however, note that, first, the control vector, as a rule, has a dimension less than the state vector; therefore, the system of nonlinear equations has many solutions with respect to control, and second, like backstepping , it is a manual method that does not lend itself to machine automation.

There is an approach to synthesis problem on the base of a Bellman equation

$$-\frac{d\mu(\mathbf{x})}{dt} = \min_{\mathbf{u}\in U}\left\{ \left(\frac{\partial \mu(\mathbf{x})}{\partial \mathbf{x}}\right)^T \mathbf{f}(\mathbf{x},\mathbf{u}) + \frac{\partial F(\mathbf{x})}{\partial \mathbf{x}}\mathbf{f}(\mathbf{x},\mathbf{u}) + f_0(\mathbf{x},\mathbf{u})\right\}. \tag{2.28}$$

If the Bellman function $\mu(\mathbf{x})$ is known, then the control function is found from the Bellman equation (2.28)

$$\mathbf{u} = \text{argmin}\left\{ \left(\frac{\partial\mu(\mathbf{x})}{\partial\mathbf{x}}\right)^T \mathbf{f}(\mathbf{x},\mathbf{u}) + \frac{\partial F(\mathbf{x})}{\partial\mathbf{x}}\mathbf{f}(\mathbf{x},\mathbf{u}) + f_0(\mathbf{x},\mathbf{u}) \right\}. \qquad (2.29)$$

To solve the synthesis problem based on Bellman equation by machine learning, it is needed to approximate the Bellman function. To apply the symbolic regression method for Bellman function we rewrite the functional taking into account all initial and terminal conditions.

$$J = \sum_{j=1}^{K}\left(\int_{0}^{t_{f,j}} f_0(\mathbf{x}(t,\mathbf{x}^{0,j}),\mathbf{u})dt + p_1\sqrt{\sum_{i=1}^{n}(x_i^f - x_i(t_{f,i},\mathbf{x}^{0,j}))^2} \right) \to \min_{\mu(\mathbf{x})}, \qquad (2.30)$$

where p_1 is a weight coefficient, and $\mathbf{x}(t,\mathbf{x}^{0,K})$ is a partial solution of the system with control (2.29) from initial condition $\mathbf{x}^{0,j}$.

There are also some other analytical methods to solve the synthesis problem, like methods of modal control [13] for linear systems, as well as synthesis based on the application of the Lyapunov function [14,15] etc., but all known analytical synthesis methods are appropriate for the specific type of model; therefore they cannot be considered universal.

Today in most applications, as a rule, specialists solve the problem of control synthesis using the so-called technical synthesis. According to the model, they determine the control channels, i.e. determine which components of the control vector affect the components of the state vector. Further, regulators are inserted into these channels, most often a PID regulator, or some other regulator, even possibly nonlinear. Then, with the help of a computer, the parameters of these regulators are found. A majority of the control systems have been built using such technical approach. In the present period, this method is also applied for robots, but this is essentially the manual labor of a developer and does not at all meet modern challenges.

Previously, automatic control systems were used only in missiles and in the autopilots of aircraft and submarines. Now robots have appeared, and the number of these robots, taking into account additive technologies, is growing catastrophically every year, and the use of technical methods for creating automatic control systems for them is the main obstacle for their development and implementation. Writing by hand a control system program for robots becomes a very difficult task. For example, how many operators will the robot control program contain, which simulates the actions of a fly? The fly controls a complex wing motion that allows it to hang motionless in the air, and it can move along a vertical surface and even with a negative slope. Further, the fly sees dangers and makes complex movements so as not to be caught. At the same time, like an ordinary animal, the fly is looking for food and the possibility of reproduction. With a simple, most optimistic estimate, the control system for such an object should contain more than a million programming operators. Probably such a program could be written by a large team of programmers. But here, and when creating even more complex control systems, the need to automate the synthesis of the control system is obvious.

2.4 Synthesized Optimal Control Problem

The formulated problems of optimal control and synthesis of control are, in fact, the main research fields in the optimal control theory. But unlike theoretical calculations, where the main criterion of quality is contained in the quality functional, practical control systems are forced to fend off a whole series of emerging difficulties associated with inaccuracies in the model used in calculations, arising noises in the state of the system, including deviations at the initial position, the need to quickly respond to changes in real conditions, be able to recalculate the trajectory of movement on board in real time, etc.

There is a field of the so-called robust systems [16, 17], in the development of which it is possible to take into account the discrepancy in the state of the object, which allows the system to remain stable during operation. However, in this case, one has to sacrifice optimality in order to lay a margin for the quality of functioning.

Ideally, we still strive to develop optimal systems that are the best in terms of the given criterion. In this case, the optimal control problem (see Sect. 2.2) is solved first. But in fact its solution cannot be directly realized on a board processor of control object since the obtained optimal control function is a function of time and its realization leads to open-loop control system, so any discrepancy in time of object movement and control will lead to the fact that the control goal will not be achieved and the value of the quality criterion will differ from that obtained during mathematical calculations. In practical control system design, the caused discrepancy between the real trajectory of the object and the obtained optimal one is compensated by the synthesis of a feedback motion stabilization system relative to the optimal trajectory.

But due to the introduction of the stabilization system, we again lose the optimality. Indeed, a number of facts indicate this:

- Construction of the stabilization system changes the mathematical model of the object and the received control might not be optimal for the new model.
- The error in the motion of the object along the trajectory can be both in time and in position. Both these errors could lead to non-optimal motion.
- The stabilization system must have control resource to return the object on the trajectory. This means that when calculating the optimal control it is necessary to take into account that not all control resources will be available. And this, as a rule, is omitted in the calculations.
- And the last but not least, that the motion of the object in the neighborhood of the programmed trajectory may differ significantly from the optimal one in terms of the value of the functional.

There is also a one more circumstance that complicates the implementation of the solution to the optimal control problem. In the classical formulation of the optimal control problem no additional requirements are put forward for the mathematical model of the control object. It follows that the problem is solved for any object, including not stable or possessing special properties, bifurcations, cycles, poles. In practical implementation, the inaccuracies of the mathematical model behave differently depending on the qualitative characteristics of the system of differential equations of the model, and they are also compensated by the feedback stabilization system.

In practice, engineers have long understood the difficulties of controlling unstable objects, so they initially make the object stable and then solve the problems of control. It is known that objects possess good properties for control, when their mathematical models are stable in the phase space. Driven by the analysis of practically implemented control systems, in this section, we would like to present a new numerical formulation of optimal control problem based on stabilization system synthesis with the main focus on its feasibility.

According to the approach, the optimal control problem is supposed to be solved after ensuring stability to the control object in the state space. Therefore, this approach is called synthesized optimal control. Its key idea is that a control function is found such that the system of differential equations will always have a stable equilibrium point in the state space. With that, the control system contains parameters that affect the position of the equilibrium point. Consequently, the object is controlled by changing the position of the equilibrium point.

Consider the Problem Statement of Synthesized Optimal Control

Given a mathematical model of the control object in the form of the system of differential equations

$$\dot{\mathbf{x}} = \mathbf{f}(\mathbf{x}, \mathbf{u}), \tag{2.31}$$

where $\mathbf{x}$ is a state space vector, $\mathbf{x} \in \mathbb{R}^n$, $\mathbf{u}$ is a control vector, $\mathbf{u} \in U \subseteq \mathbb{R}^m$, U is a limited compact set, $m \leq n$.

The initial condition is given by

$$\mathbf{x}(0) = \mathbf{x}^0. \tag{2.32}$$

Given the terminal condition

$$\mathbf{x}(t_f) = \mathbf{x}^f, \tag{2.33}$$

where t_f is the time of achieving the terminal condition, t_f is not given, but limited

$$t_f \leq t^+, \tag{2.34}$$

and t^+ is given.

The quality criterion is given by

$$J_1 = \int_0^{t_f} f_0(\mathbf{x}, \mathbf{u}) dt \to \min_{\mathbf{u} \in U}. \tag{2.35}$$

It is necessary to find a control in the following form:

$$\mathbf{u} = \mathbf{h}(\mathbf{x}^*(t) - \mathbf{x}) \in U, \tag{2.36}$$

where $\mathbf{x}^*(t)$ is a function of time.

The function

$$\mathbf{h}(\mathbf{x}^*(t) - \mathbf{x}) : \mathbb{R}^n \to \mathbb{R}^m \tag{2.37}$$

is searched such that it possesses a feasibility property [18], i.e. for any time $t = t_k \leq t_f$ the system

$$\dot{\mathbf{x}} = \mathbf{f}(\mathbf{x}, \mathbf{h}(\mathbf{x}^*(t_k) - \mathbf{x})) \tag{2.38}$$

has a stable equilibrium point

$$\tilde{\mathbf{x}}(\mathbf{x}^*(t_k)) \in \mathbb{R}^n, \tag{2.39}$$

$$\mathbf{f}(\tilde{\mathbf{x}}, \mathbf{h}(\mathbf{x}^*(t_k) - \tilde{\mathbf{x}})) = 0, \tag{2.40}$$

$$\det(\mathbf{A} - \lambda \mathbf{E}) = \lambda^n + a_{n-1}\lambda^{n-1} + \ldots + a_1\lambda + a_0 = \prod_{j=1}^n (\lambda - \lambda_j) = 0, \tag{2.41}$$

where

$$\lambda_j = \alpha_j + i\beta_j, \ \alpha_j < 0, \ j = 1, \ldots, n, \tag{2.42}$$

$i = \sqrt{-1}$,

$$\mathbf{A} = \frac{\partial \mathbf{f}(\tilde{\mathbf{x}}, \mathbf{h}(\mathbf{x}^*(t_k) - \tilde{\mathbf{x}}))}{\partial \mathbf{x}}. \tag{2.43}$$

Algorithmically, the solution of the synthesized optimal control problem and finding the control function (2.36) is assumed to be carried out in two stages as two sequential tasks.

1st Stage: Stabilization System Synthesis

Initially, on the stabilization stage, the control synthesis problem is solved to provide existence of the stable equilibrium point in the state space. Consider its problem statement to be solved numerically by some machine learning technique.

The mathematical model of the control object (2.31) is given.

The set of initial conditions is given by

$$X_0 = \{\mathbf{x}^{0,1}, \ldots, \mathbf{x}^{0,K}\}. \tag{2.44}$$

The terminal position is given. This can be any point relative to which the system will be stabilized. Its position cannot coincide with the terminal condition (2.33) in the optimal control problem.

$$\mathbf{x}(t^*) = \mathbf{x}^* \in \mathbb{R}^n, \tag{2.45}$$

where t^* is not given, but limited

$$t^* = \begin{cases} t, \text{ if } t < t^+ \text{ and } \|\mathbf{x}^* - \mathbf{x}(t, \mathbf{x}^0)\| \le \varepsilon \\ t^+, \text{ otherwise} \end{cases}, \tag{2.46}$$

where $\mathbf{x}(t, \mathbf{x}^0)$ is a partial solution of the system (2.31), and ε and t^+ are given positive values.

It is necessary to find the control function in the form

$$\mathbf{u} = \mathbf{h}(\mathbf{x}^* - \mathbf{x}) \tag{2.47}$$

that partial solutions of the system of differential equations

$$\dot{\mathbf{x}} = \mathbf{f}(\mathbf{x}, \mathbf{h}(\mathbf{x}^* - \mathbf{x})) \tag{2.48}$$

from any initial condition from the area (2.44)

$$\mathbf{x}^{0,i} \in X_0, \ i = 1, \dots, K \tag{2.49}$$

will achieve terminal condition (2.45) with an optimal value of the following criterion:

$$J_2 = \sum_{i=1}^{K} \left(t_i^* + p_1 \|\mathbf{x}^* - \mathbf{x}(t_i^*, \mathbf{x}^{0,i})\| \right), \tag{2.50}$$

where

$$t_i^* = \begin{cases} t, \text{ if } t < t^+ \text{ and } \|\mathbf{x}^* - \mathbf{x}(t, \mathbf{x}^{0,i})\| \le \varepsilon_1 \\ t_1^+, \text{ otherwise} \end{cases}, \tag{2.51}$$

$$\|\mathbf{x}^* - \mathbf{x}(t, \mathbf{x}^{0,i})\| = \sqrt{\sum_{i=1}^{n} (\mathbf{x}^* - \mathbf{x}(t, \mathbf{x}^{0,i}))^2}, \tag{2.52}$$

p_1 is a weight coefficient, and ε and t_1^+ are given positive values.

2nd Stage: Solution of the Optimal Control Problem

At the second stage of the synthesized optimal control, after solution of the control synthesis problem (2.31)–(2.52), the optimal control problem (2.31)–(2.35) is solved for the mathematical model (2.48), where it is necessary to find a control function in the following form:

$$\mathbf{x}^*(t) = \mathbf{v}(t) \tag{2.53}$$

in order to minimize the given criterion (2.35).

Note that at the second stage the searched function (2.53) has the same dimension as the state space. As a particular case, it can be searched as a piece-constant function,

$$\mathbf{v}(t) = \mathbf{x}^{*,i}, \quad \text{if } (i-1)\Delta \le t < i\Delta, \tag{2.54}$$

where $\mathbf{x}^{*,i}$ are found optimal values of the equilibrium point coordinates, $i = 1, \dots, K$, Δ is a given time interval,

$$K = \left\lfloor \frac{t^+}{\Delta} \right\rfloor. \tag{2.55}$$

Thus, according to the method of synthesized optimal control, the stability of the object is first ensured, i.e. an equilibrium point appears in the phase space. In the neighborhood of the equilibrium point, the phase trajectories contract, and this property determines the feasibility of the system. This is the main strength of this approach in comparison with the problem of optimal control, described in Sect. 2.2, where as a result the developer gets an open-loop control system.

To provide this property, it is necessary to numerically solve the problem of synthesizing the stabilization system in order to obtain expressions for the control and substitute them in the right-hand sides of the object model. A logical question may arise here: if it is still necessary to solve the problem of synthesis, may it be better to consider the problem directly as a problem of general synthesis (as described in Sect. 2.3)? As noted earlier, the problem of general synthesis is very difficult, including from a computational point of view. It must be solved in advance, taking into account all possible phase constraints. However, in most application systems, this may not always be possible. Typically, the control object operates in a dynamic environment and it is extremely important that the optimal path can be calculated on board. The approach based on synthesized optimal control makes it possible to do just that. The problem of the stabilization system synthesis is solved in advance, and already the optimal position of the equilibrium points, as parameters of the control system, can be calculated either in advance or in real time on board.

2.5 Model Identification Problem

Usually, when creating a control system, a mathematical model of the control object is required, as a rule, in the form of a system of differential equations. When the model is obtained, researchers or developers study it and calculate the optimal control system using one of the well-known methods. Receiving a mathematical model for the control object is a complex manual process. Note that any new movable element of the control object changes its mathematical model and increases the number of generalized coordinates and control channels. So this process should also be automated and, therefore, must be considered as the task for machine learning.

In the identification problem, the mathematical model of the control object is not fully or partially known, but the researcher has a real control object or its physical simulator. In this case, the real control object or physical simulator layout is an unknown function. The space of the input vectors of this function is the space of admissible controls for this object.

Assume that it is known how many components have a control vector $\mathbf{u} = [u_1 \ldots u_m]^T$ and a state space vector $\mathbf{x} = [x_1 \ldots x_n]^T$ of the control object. A time interval Δt is set.

A control function of time is set in each interval

$$\mathbf{u}(t) = \mathbf{v}^k(t), \quad (k-1)\Delta t \le t < k\Delta t, \tag{2.56}$$

where $\mathbf{v}^k(t)$ are given values of the function of time, $k = 1, \ldots, N$.

The control function (2.56) is supplied to the real object, which changes its state depending on its current state, $\mathbf{x}(t, \mathbf{x}^0)$, where $\mathbf{x}^0$ is a state of the real control object in initial moment of time.

Define M points on the axis of time $t_0, t_1, \ldots, t_{M-1}$, and values of the state vector are saved in these points

$$X = \{\mathbf{x}^0, \ldots, \mathbf{x}^{M-1}\}, \tag{2.57}$$

where $\mathbf{x}^j = \mathbf{x}(t_j, \mathbf{x}^0)$, $j = 0, \ldots, M-1$.

Setting the control function (2.56) and storing points of the state vector (2.57) are repeated a given number L of times.

In the result, the sets of control function and points of the state space vector are obtained

$$\langle U, X \rangle, \tag{2.58}$$

where

$$U = \{\mathbf{u}^1(\cdot), \ldots, \mathbf{u}^L(\cdot)\}, \tag{2.59}$$

$$X = \{X_1, \ldots, X_L\}, \tag{2.60}$$

$$X_j = \{\mathbf{x}^{j,1}, \ldots, \mathbf{x}^{j,M}\}, \quad j = 1, \ldots, L. \tag{2.61}$$

The pair (2.58) is called a training set.

It is necessary to find a system of ordinary differential equations in the form

$$\dot{\tilde{\mathbf{x}}} = \tilde{\mathbf{f}}(\tilde{\mathbf{x}}, \tilde{\mathbf{u}}), \tag{2.62}$$

where $\tilde{\mathbf{x}} = [\tilde{x}_1 \ldots \tilde{x}_n]^T$, $\tilde{\mathbf{u}} = [\tilde{u}_1 \ldots \tilde{u}_m]^T$.

Let

$$\tilde{\mathbf{x}}(t, \mathbf{x}^0, \mathbf{u}^j(\cdot)), \quad j = 1, \ldots, L, \tag{2.63}$$

be a value of the partial solution of the differential equation system (2.62) with a control function $\mathbf{u}^j(\cdot)$, from initial conditions $\mathbf{x}^0$ in the moment of time t.

The right part function $\mathbf{f}(\tilde{\mathbf{x}}, \tilde{\mathbf{u}})$ is searched on the criterion

$$J = \sqrt{\sum_{j=1}^{L} \sum_{i=0}^{M-1} \sum_{k=1}^{n} (\tilde{x}_k(t_i, \mathbf{x}^{j,0}, \mathbf{u}^j(\cdot)) - x_k^{j,i})^2} \to \min_{\tilde{\mathbf{f}}(\tilde{\mathbf{x}}, \mathbf{u})} . \tag{2.64}$$

Thus, machine learning problem of model identification is to find the function (2.62) by minimizing the functional (2.64).

References

1. Betts, J.T.: Survey of numerical methods for trajectory optimization. J. Guid. Control. Dyn. **21**, 193–207 (1998)
2. Gill, P.E., Murray, W., Wright, M.H.: Practical Optimization. Academic, Cambridge, MA (1981)
3. Pontryagin, L.S., Boltyanskii, V.G., Gamkrelidze, R.V., Mishchenko, E.F.: The Mathematical Theory of Optimal Process. Gordon and Breach Science Publishers, New York/London/Paris/Montreux/Tokyo (1985)
4. Bellman, R., Glickberg, I., Gross, O.: Some Aspects of the Mathematical Theory of Control Processes. Rand Corporation, Santa Monica (1958)
5. Bellman, R.E, Kalaba, R.E.: Dynamic Programming and Modern Control Theory. Academic, New York/London (1966)
6. Bellman, R.E., Dreyfus, S.E.: Applied Dynamic Programming. Princeton University Press, Princeton (1962)
7. Letov, A.M.: Analytical design of controllers. J. Automatica i Telechanika **21**(4), 436–441 (1960)
8. Boltyanskii, V.G.: Mathematical Methods of Optimal Control. Holt, Rinehart and Winston, New York (1971)
9. Kokotovic, P.V.: The joy of feedback: nonlinear and adaptive. IEEE Control Syst. Mag. **12**(3), 7–17 (1992)
10. Khalil, H.K.: Nonlinear Systems. Prentice Hall, New York (2002)
11. Kolesnikov, A.A.: The ADAR method and theory of optimal control in the problems of synthesis of nonlinear control systems. Mechatron. Autom. Control **18**(9), 579–589 (2017)
12. Podvalny, S.L., Vasiljev, E.M.: Analytical synthesis of aggregated regulators for unmanned aerial vehicles. J. Math. Sci. **239**, 135–145 (2019)
13. Simon, J.D., Mitter, S.K.: A theory of modal control. Inf. Control. **13**, 316–353 (1968)
14. Clarke, F.: Lyapunov functions and feedback in nonlinear control. In: de Queiroz, M., et al. (eds.) Optimal Control, Stabilization and Nonsmooth Analysis. LNCIS 301, pp. 267–282. Springer, Berlin/Heidelberg (2009)
15. Agarwal, R., O'Regan, D., Hristova, S.: Stability by Lyapunov like functions of nonlinear differential equations with non-instantaneous impulses. J. Appl. Math. Comput. **53**, 147–168 (2017)
16. Dullerud, G.E., Paganini, F.: A Course in Robust Control Theory: A Convex Approach. Springer, New York (2000)
17. Calafiore, G., Dabbene, F. (eds.): Probabilistic and Randomized Methods for Design Under Uncertainty. Springer, London (2006)
18. Diveev, A., Shmalko, E.: Multi-point Stabilization Approach to the Optimal Control Problem with Uncertainties (2020). https://doi.org/10.1007/978-3-030-65739-0_10

Chapter 3
Numerical Solution of Machine Learning Control Problems

Abstract This chapter discusses general issues in the numerical solution of machine learning control problems. As parametric machine learning approach, the most popular and widespread apparatus of neural networks is considered. Theoretical substantiations are given for the general possibility of using machine learning methods for searching functions, namely the Kolmogorov–Arnold theorem. The only general approach of structural-parametric search of functions based on the methods of symbolic regression is presented. To overcome computational difficulties, it is proposed to use the principle of small variations. A description of the genetic algorithm is given as the main search mechanism in the space of structures, and in addition, it can also be used to adjust the parameters of a given structure of a function in parametric search.

3.1 Artificial Neural Networks

According to our definition, machine learning task is searching for unknown functions. The most general numerical approach to the approximation of any function, including those with discontinuities of the first kind, is the approximation by various polynomial series, for example, Taylor, Fourier, etc. But in the case of multidimensional functions of a vector argument, the use of such series is difficult.

Today, a neural network is considered a universal approximator [1]. In fact, any neural network is a sequential set of linear transformations, which generally corresponds to the first term of a multidimensional Taylor series expansion.

The problem of finding an unknown function using neural networks is of a parametric type, when the required function is specified by the researcher with an accuracy up to parameters, and the machine is looking for these parameters.

Indeed, any type of neural networks is a function with a given structure and a large number of parameters. The structure is a sequence of linear vector transformations. Each transformation is performed on a separate layer. The nonlinearities

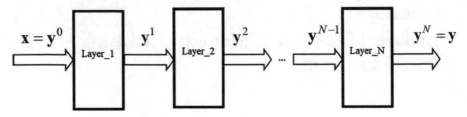

Fig. 3.1 General scheme of a multilayer perceptron

introduced between the layers are added in order to separate some linear transformations from others.

For example, on some layer, a linear transformation of the input vector $\mathbf{x}$ is performed

$$\mathbf{y} = \mathbf{A}\mathbf{x}.$$

On the next layer, the linear transformation is again carried out

$$\mathbf{z} = \mathbf{B}\mathbf{y}.$$

So, if the layers are not separated by some nonlinearity, then in fact one linear transformation takes place

$$\mathbf{z} = \mathbf{B}\mathbf{A}\mathbf{x}.$$

In order to separate these linear transformations, nonlinearities are introduced between them. The type of nonlinearity is also set by the researcher at his discretion. Previously, a sigmoid function [2, 3] was used as the main activation function, and now ReLU is very popular [4].

Let us consider the transformations of the input vector in the neural network using the example of a multilayer perceptron with N layers, as shown in Fig. 3.1.

$$\mathbf{z}^i = \mathbf{A}^i \mathbf{y}^{i-1} + \mathbf{b}^i, \quad i = 1, \ldots, N \tag{3.1}$$

$$\mathbf{y}^i = F(\mathbf{z}^i), \tag{3.2}$$

where $\mathbf{b}^i$ is a vector of displacement, in some NN it may be equal to zero, and F is a componentwise nonlinear transformation of the components of the vector $\mathbf{z}^i$,

$$F(\mathbf{z}^i) = \begin{bmatrix} f_1(z_1) \\ f_2(z_2) \\ \cdots \\ f_N(z_N) \end{bmatrix},$$

where f_i are nonlinear transformations and are called activation functions.

During NN training, the components of the matrix $\mathbf{A}^i$ and the vector $\mathbf{b}^i$ are searched in the function (3.1).

As can be seen, a neural network is a function with a given structure. You can change the structure by increasing or decreasing the number of layers, or change the activation functions (3.2), but these actions are rarely performed by developers when training neural networks.

Similar linear transformations can be written for other types of neural networks, such as the Elman and Jordan recurrent networks [5], the Hopfield feedback network [6], etc.

The possibility of approximating any function using a neural network is based on the Kolmogorov–Arnold representation theorem [7, 8] in that any multidimensional function can be represented as a finite composition of continuous functions of one variable. Its extension to ANN is also known as Kolmogorov–Arnold—Hecht-Nielsen theorem [9].

Today, artificial neural networks have developed into a powerful direction with a good software and application base. As a result, the neural network gives us a representation of the function, implements its operation, but note that it does not provide any information about this function, and the function can be exponential or polynomial, have kinks and discontinuities, etc. ANN performs substitution of this function but does not disclose its type and properties. And this is very important from the point of view of understanding the ongoing processes, especially in the field of control.

Another approach to machine search for an unknown function is the use of symbolic regression methods. These methods make it possible to simultaneously carry out a structural-parametric search for a function and, as a result, give a description of the function that reflects the properties of this function. In addition, in the general case, any symbolic regression method can approximate a neural network.

3.2 General Approach of Symbolic Regression

In the modern conception and in the form in which these conceptions are used in this book, symbolic regression methods are methods for encoding mathematical expressions and a collection of algorithms for finding the optimal mathematical expressions in the space of these codes in machine learning problems (Fig. 3.2).

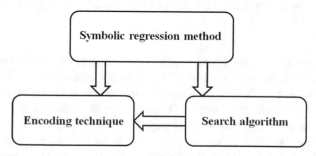

Fig. 3.2 General scheme of symbolic regression methods

Encoding Technique

To encode a mathematical expression using any symbolic regression method, it is necessary to create an initial alphabet of primitives. It should include a set of elementary functions from which the mathematical expression is constructed and a set of arguments of this mathematical expression. The most general and convenient way for coding and decoding is to split the set of elementary functions into subsets of functions with a certain number of arguments.

$$
\begin{aligned}
F_1 &= \{f_{1,1}(z), f_{1,2}(z), \ldots, f_{1,v_1}(z)\}, \\
F_2 &= \{f_{2,1}(z_1, z_2), f_{2,2}(z_1, z_2), \ldots, f_{2,v_2}(z_1, z_2)\}, \\
F_3 &= \{f_{3,1}(z_1, z_2, z_3), f_{3,2}(z_1, z_2, z_3), \ldots, f_{3,v_3}(z_1, z_2, z_3)\}, \\
&\ldots
\end{aligned}
\tag{3.3}
$$

where v_i is the number of elements in the set F_i.

Here the subscript of the set identifier and the first subscript of the element indicate the number of arguments of the function. In most cases, it is sufficient to use functions with one and two arguments. Note that functions with three arguments are convenient to describe IF-operator. For example,

$$
f_{3,1}(z_1, z_2, z_3) = \begin{cases} z_2, & \text{if } z_1 \leq 0 \\ z_3, & \text{otherwise} \end{cases}.
\tag{3.4}
$$

In the case of representing the alphabet of primitives in the form (3.3), the set of arguments of the mathematical expression corresponds to the set of functions with zero arguments

$$
F_0 = \{f_{0,1}, f_{0,2}, \ldots, f_{0,v_0}\}.
\tag{3.5}
$$

Representation of alphabet primitives in the form (3.3), (3.5) is not mandatory. In some methods of symbolic regression, all primitive functions, regardless of the number of arguments, are placed in one set. In any case, when decoding a mathematical expression, and this is always required when calculating the value of the desired mathematical expression, the number of arguments of the desired function is important information.

Note that when representing the alphabet in the form (3.3), any function, including the argument of the mathematical expression, is identified by an integer vector of two components

$$
f_{a,b} \leftrightarrow [a\ b]^T.
\tag{3.6}
$$

To solve machine learning control problems and search for unknown functions, it is necessary to determine the rules for composing mathematical expressions from elementary functions. Various methods are also possible here. For example, you can define the structure of a function. Then change the elements in it, keeping the first index. The most universal way of composing mathematical expressions from elementary functions is to represent a sequence of elementary functions as a composition of these elementary functions nested into each other. For example,

$$f_{a_1,b_1} f_{a_2,b_2} f_{a_3,b_3} = f_{a_1,b_1} \circ f_{a_2,b_2} \circ f_{a_3,b_3} = f_{a_1,b_1}\left(f_{a_2,b_2}\left(f_{a_3,b_3}(\ldots)\right)\right). \tag{3.7}$$

Note that such a representation in symbolic regression methods is fully consistent with the Kolmogorov–Arnold theorem on the representation of functions. According to it, if f is a multidimensional continuous function, then f can be written as a finite composition of continuous functions of one variable and a binary operation of addition. In terms of symbolic regression alphabet (3.3), functions with one variable are accumulated in the set F_1, and their combinations (addition and multiplication) in the set F_2.

Obviously, when forming compositions, a simple sequential listing of functions can lead to violation of the mathematical notation. The rules to form compositions for obtaining the correct mathematical notations are defined in different methods of symbolic regression.

Here are the rules to form correct compositions from function codes constructed for the case of dividing functions into subsets (3.3) taking into account the number of arguments.

Suppose a sequence of functions is given by

$$S = f_{a_1,b_1} \circ f_{a_2,b_2} \circ \ldots \circ f_{a_i,b_i} \circ \ldots \circ f_{a_K,b_K}. \tag{3.8}$$

To determine the correctness of code record, let us introduce the concept of the composition item index.

Definition 3.1. A `composition item index` indicates the minimum number of items that must follow the given item.

The index of the f_{a_1,b_i} item in the composition record is calculated by the formula

$$T(i) = T(i-1) + a_i - 1 = 1 - i + \sum_{j=1}^{i} a_j, \ 1 \leq i \leq K. \tag{3.9}$$

Definition 3.2. The composition record is `correct` if all elements of the record, except the last one, have a positive index, and the index of the last item is equal to zero.

$$\begin{aligned} T(i) &> 0, \ i = 1,\ldots,K-1, \\ T(K) &= 0. \end{aligned} \tag{3.10}$$

The set of elementary functions must be reachable in order to be used in symbolic regression methods.

Definition 3.3. A set of elementary functions possesses `reachability` property if, for any given bounded numbers A and B and a given small value ε, one can construct a finite composition of functions from this set

$$f_{a_1,b_1} \circ \ldots \circ f_{a_k,b_k} \tag{3.11}$$

1. Initialization:
- randomly generate the set of possible solutions

2. Evaluation:
- Determine functional values of possible solutions

3. Until convergence repeat:

   3.1. Selection:
   - Select "parents" from the set of possible solutions

   3.2. Crossover:
   - Randomly exchange parts of parents and generate new possible solutions

   3.3. Mutation:
   - Randomly change some parts of new possible solutions

   3.4. Evaluation:
   - Determine functional values for the new set of possible solutions

Fig. 3.3 GA scheme

such that the inequality holds

$$|f_{a_1,b_1}(f_{a_2,b_2}\cdots f_{a_K,b_K}(A)\ldots) - B| \le \varepsilon. \tag{3.12}$$

So, we examined the general principles of coding mathematical expressions, with the help of which the search space is determined in symbolic regression methods. Machine learning using symbolic regression methods consists of searching for the structure of an unknown function and its parameters.

Search Algorithm

The main search engine is a genetic algorithm (GA), which is used both to search for the structure of the mathematical expression and its parameters.

The genetic algorithm for many years of its existence [10–14] has confirmed its effectiveness and does not need additional announcement.

Let us recall only the general scheme of GA for readers who have not yet had time to get acquainted with genetic algorithms (see Fig. 3.3).

One of the important features of a genetic algorithm is its ability to operate on the space of codes. Search in the code space, in contrast to the search in a vector numeric space, is complicated by the fact that the metric on the code space differs from the metric of the space in which the objective functional is calculated.

To illustrate this, consider a nonlinear programming problem

$$f(\mathbf{q}) \rightarrow \min_{\mathbf{q}}, \qquad (3.13)$$

where $\mathbf{q} \in [\mathbb{R}]^p$, $f(\mathbf{q}) : \mathbb{R}^p \rightarrow \mathbb{R}^1$.

The continuous objective function (3.13) satisfies the relations

$$|f(\mathbf{q}^1) - f(\mathbf{q}^2)| = \delta \approx 0, \quad \text{if } \|\mathbf{q}^1 - \mathbf{q}^2\| \approx 0. \qquad (3.14)$$

Let $G(\mathbf{q}) = \mathbf{g}$ be the operation of translating the real vector $\mathbf{q}$ into Gray code $\mathbf{g}$, $\mathbf{g} = [g_1 \dots g_{pc}]^T$, where c is the amount of bits allocated to encode one component of the vector $\mathbf{q}$ and p is the amount of components, $g_i \in \{0, 1\}$.

Consider the Hamming metric [15] in the code space that determines the distance between two codes as the number of mismatched bits

$$d_H(\mathbf{g}^1, \mathbf{g}^2) = \sum_{i=1}^{pc} |g_i^1 - g_i^2|. \qquad (3.15)$$

As a result, the smallest distance between the codes of two vectors is equal to one.

Let $\mathbf{g}^1 = G(\mathbf{q}^1)$ and $\mathbf{g}^2 = G(\mathbf{q}^2)$.

Then, if

$$d_H(\mathbf{g}^1, \mathbf{g}^2) = 1, \qquad (3.16)$$

this does not mean that the values of the objective functional for these vectors are close

$$f(\mathbf{q}^1) \not\approx f(\mathbf{q}^2). \qquad (3.17)$$

This happens because the calculation of the objective functional is performed using the vector components from the metric space of real vectors.

Definition 3.4. An optimization problem in which the search for a solution is performed on the code space, and the calculation of the objective functional is carried out in a metric vector space, is called a non-numerical optimization problem.

The genetic algorithm with its unique structure is capable of searching in non-numerical space. The most important feature of the genetic algorithm is that it does not use arithmetic operations to obtain new possible solutions. This feature allows to use GAs for non-numerical optimization in machine learning control problems.

The classical GA works, as a rule, with Gray codes, so vectors from the real space are translated into the space of Gray codes.

Here is a small computational life hack. In case of one-point crossover, it is sufficient to translate into the Gray code and back only one component from the vectors selected for crossover, because the other components do not undergo changes as a result of crossover.

Suppose two real vectors are selected for crossover

$$\begin{aligned}
\mathbf{q}^\alpha &= [q_1^\alpha \dots \mathbf{q}_p^\alpha]^T, \\
\mathbf{q}^\beta &= [q_1^\beta \dots \mathbf{q}_p^\beta]^T.
\end{aligned} \qquad (3.18)$$

Randomly determine the component in which the crossover point will be located. For this purpose, choose a random integer

$$k \in \{1, \ldots, p\}. \tag{3.19}$$

Convert components k in the selected vectors into the Gray code

$$\begin{aligned}
\mathbf{g}^{\alpha,k} &= G(q_k^\alpha) = [g_1^{\alpha,k} \ldots g_c^{\alpha,k}]^T, \\
\mathbf{g}^{\beta,k} &= G(q_k^\beta) = [g_1^{\beta,k} \ldots g_c^{\beta,k}]^T.
\end{aligned} \tag{3.20}$$

Find randomly the crossover point for the Gray codes

$$r \in \{1, \ldots, c\}. \tag{3.21}$$

Produce crossover on the Gray codes of the components of the selected vectors

$$\begin{aligned}
\mathbf{g}^{\gamma,k} &= [g_1^{\alpha,k} \ldots g_r^{\alpha,k} g_{r+1}^{\beta,k} \ldots g_c^{\beta,k}]^T, \\
\mathbf{g}^{\delta,k} &= [g_1^{\beta,k} \ldots g_r^{\beta,k} g_{r+1}^{\alpha,k} \ldots g_c^{\alpha,k}]^T.
\end{aligned} \tag{3.22}$$

Translate the new received Gray codes into numbers

$$\begin{aligned}
q_k^\gamma &= G^{-1}(\mathbf{g}^{\gamma,k}), \\
q_k^\delta &= G^{-1}(\mathbf{g}^{\delta,k}),
\end{aligned} \tag{3.23}$$

where $G^{-1}(\mathbf{g})$ is the reverse translation function from the Gray code to the number. Exchange the remaining components of the vectors selected for crossover.

In result, we get two new numeric vectors

$$\begin{aligned}
\mathbf{q}^\gamma &= [q_1^\alpha \ldots q_{k-1}^\alpha \, q_k^\gamma \, q_{k+1}^\beta \ldots q_c^\beta]^T, \\
\mathbf{q}^\delta &= [q_1^\beta \ldots q_{k-1}^\beta \, q_k^\delta \, q_{k+1}^\alpha \ldots q_c^\alpha]^T.
\end{aligned} \tag{3.24}$$

This type of crossover is more economical and more expedient than transferring all the components of the real vector to the Gray code and vice versa, since the components that did not hit the crossover point are not changed.

Thus, to summarize, all symbolic regression methods that are used for machine learning control encode possible solutions (i.e. mathematical expressions of the unknown functions) and look for an optimal solution in the space of these codes, while the estimate of a possible solution is calculated in the space of real functions.

All methods of symbolic regression use the genetic algorithm to find the optimal solution. Since the symbolic regression methods differ depending on the form of encoding the mathematical expression, they also differ in the basic GA operations of crossover and mutation applied to these codes. Thus, to create new methods of symbolic regression, it is necessary to determine the form of encoding possible solutions and redefine the operations of crossover and mutation in the genetic algorithm so that as a result of these operations, the correct codes of new possible solutions are obtained.

Genetic algorithms used in well-known symbolic regression methods are considered in Chap. 4. The main features of the genetic algorithm for simultaneous structural and parametric multicriterial search are considered in detail in Sect. 3.4 of this chapter. In the next Sect. 3.3, we touch upon the issue of the complexity of the search on non-numeric spaces of structures and consider the principle of small variations of the basic solution, as one of the promising ways to overcome these difficulties.

3.3 The Principle of Small Variations of the Basic Solution

As shown in the previous section, searching for an optimal solution in the space of codes is complicated by the fact that this task belongs to the class of non-numerical optimization problems. For such search spaces it is impossible to use evolutionary algorithms with arithmetic operations. Most of the known evolutionary algorithms include arithmetic operations to transform possible solutions and produce evolution. Therefore, genetic algorithm is a main searching algorithm on the space of codes that does not use arithmetic operations in its steps. At the same time, with certain complex forms of coding in different symbolic regression methods, the construction of new operations of crossover and mutation is a significant problem. Studies of this problem have led to the formulation of the principle of small variations of the basic solution [16].

Consider a universal approach to the construction of genetic algorithms for solving non-numerical optimization problems, based on the application of the principle of small variations of the basic solution.

The essence of this approach is as follows. One possible solution is coded, which is called a basic solution. In complex problems, it is very advisable to use such basic solution that can be close to the optimal one in the opinion of the researcher or developer in order to speed up the search process. Next, small variations of this code are defined such that any of them varies the code so that as a result the correct code of the new possible solution is obtained. All small variations are coded. As far as the small variation is considered as an operator acting in the space of codes of the basic solution, therefore, a code of the small variation itself in all cases is an integer vector containing information necessary to perform actions on the code according to the small variation operator.

Let us have a detailed view on the principle of small variations of the basic solution.

Consider a code space Ξ^n. Any element of this space is a code vector of some non-numerical construction, in particular a mathematical expression.

$$\mathbf{y} = [y_1 \ldots y_n]^T. \tag{3.25}$$

Any element of this code has a meaning from a set of possible characters.

$$y_i \in A = \{0, a_1, \ldots, a_L\}. \tag{3.26}$$

In some cases, the character for the item code can be an integer. Then the code of a non-numerical element will be a set of integers, for which arithmetic operations are inapplicable, as, for example, for postal codes.

Definition 3.5. An `elementary variation` of the code of a non-numerical element is to replace the value of the code element with another value from a set of possible characters.

Replacing one character with another may not always result in a new correct code that matches some new non-numerical construction.

Definition 3.6. A `small variation` of the non-numerical construction code is a minimum set of elementary variations that allows to obtain a new valid non-numerical construction code.

Define a set of small variations for the given code space

$$\Omega(\Xi^n) = \{\delta_1(\mathbf{y}), \ldots, \delta_M(\mathbf{y})\}. \tag{3.27}$$

There may be several small variations depending on the code.

Definition 3.7. A set of small variations is `complete` if it is possible to derive from any valid code of a non-numerical construction any other valid code from the code space.

Now introduce the concept of a distance in the code space.

Definition 3.8. A `distance` between two elements $\mathbf{y}^1$, $\mathbf{y}^2$ of the code space Ξ^n is the minimum number of small variations of one code $\mathbf{y}^1$ required to get another code $\mathbf{y}^2$

$$\|\mathbf{y}^1 - \mathbf{y}^2\|_\Xi = d, \tag{3.28}$$

where

$$d = \min_r \{\mathbf{y}^2 = \delta_{k_1}(\ldots \delta_{k_r}(\mathbf{y}^1)\ldots)\}. \tag{3.29}$$

Definition 3.9. Δ-`neighborhood` of the code $\Delta(\tilde{\mathbf{y}})$ is a subset of all codes that are at a distance of no more than Δ from the code $\tilde{\mathbf{y}}$

$$\forall \mathbf{y} \in \Delta(\tilde{\mathbf{y}}) \Rightarrow \|\tilde{\mathbf{y}} - \mathbf{y}\|_\Xi \leq \Delta. \tag{3.30}$$

To encode a small variation, an integer vector is introduced

$$\mathbf{w} = [w_1 \ldots w_r]^T, \tag{3.31}$$

where r is the dimension of the vector of variations, which depends on the form of encoding, w_1 is the number or type of the small variation, the remaining components determine the number of the variable elements in the code, and w_r is, as a rule, a new value of the variable element.

GA based on the principle of small variations of the basic solution contains the following steps:

1. Establish the basic solution, which, according to the researcher, is the closest to the possible optimal solution.

$$\mathbf{y}^0 = [y_1^0 \ldots y_n^0]^T. \tag{3.32}$$

In the practical problems of control synthesis, discussed in Chap. 5, a proportional controller or a linear transformation of the state vector was most often used as the basic solution for various control objects.

2. Generate initial population in the form of ordered multisets of variation vectors

$$\mathbf{W}^i = (\mathbf{w}^{i,1}, \ldots, \mathbf{w}^{i,d}), \ i = 1, \ldots, H, \tag{3.33}$$

where H is the number of possible solutions in the initial population and d is the number of variation vectors in one set.

Every possible solution from the initial population is obtained by applying small variations to the basic solution

$$\mathbf{y}^i = \mathbf{w}^{i,d} \circ \ldots \circ \mathbf{w}^{i,1} \circ \mathbf{y}^0. \tag{3.34}$$

From Definitions 3.9 and (3.34) it follows that every possible solution in the population belongs to the d-neighborhood of the basic solution

$$\mathbf{y}^i \in d(\mathbf{y}^0), \ i = 1, \ldots, H. \tag{3.35}$$

3. Calculate the value of the objective functional for each possible solution in the population

$$F_i = J(\Omega(\mathbf{y}^i)), \ i = 1, \ldots, H, \tag{3.36}$$

where $\Omega(\mathbf{y})$ is the function of converting the code of the non-numeric construction into a real function.

4. Until the stop condition is satisfied, the evolution cycle is implemented:

   a. Randomly select two sets of vectors of variations

$$\begin{aligned} \mathbf{W}_\alpha &= (\mathbf{w}^{\alpha,1}, \ldots, \mathbf{w}^{\alpha,d}), \\ \mathbf{W}_\beta &= (\mathbf{w}^{\beta,1}, \ldots, \mathbf{w}^{\beta,d}). \end{aligned} \tag{3.37}$$

   b. Calculate the probability of crossover by the values of the objective functional for the selected vectors

$$Pr_c = \max \left\{ \frac{F_{j_-}}{F_\alpha}, \frac{F_{j_-}}{F_\beta} \right\}. \tag{3.38}$$

If the random number generator produced a number less than Pr_c, then crossover is performed.

Randomly find the crossover point

$$c \in \{1, \ldots, d\}. \tag{3.39}$$

Exchange the vectors of variations in the selected sets after the crossover point and obtain two new sets of vectors of variations that correspond to two new solutions from the d-neighborhood of the basic solution

$$
\begin{aligned}
W_{H+1} &= (\mathbf{w}^{\alpha,1}, \ldots, \mathbf{w}^{\alpha,c}, \mathbf{w}^{\beta,c+1}, \ldots, \mathbf{w}^{\beta,d}), \\
W_{H+2} &= (\mathbf{w}^{\beta,1}, \ldots, \mathbf{w}^{\beta,c}, \mathbf{w}^{\alpha,c+1}, \ldots, \mathbf{w}^{\alpha,d}).
\end{aligned} \tag{3.40}
$$

c. Perform the mutation operation with a given probability for the obtained new possible solutions in the form of sets of vectors of variations (3.40). Randomly find the mutation point and generate a new vector of variations in this position.
d. Calculate the values of the objective functional for the obtained new possible solutions and determine, from the values of these functionals, the fate of each new possible solution, either it is discarded or included in the population instead of the currently worst possible solution.

Thus, the genetic algorithm based on the principle of small variations of the basic solution includes the same actions as the usual genetic algorithm. The crossover in it is performed in the usual way, by exchanging tail elements after the crossover point. This algorithm can be supplemented with one more loop for changing the basic solution. After performing a given number of iterations of constructing new possible solutions, the basic solution is replaced by a possible solution selected for the new basis, which is the best in terms of functional.

The principle of small variations in the basic solution was first used in the network operator method. Further, this principle was applied in other methods of symbolic regression. In all cases of its application, the method of symbolic regression in the search for a mathematical expression worked significantly better than the same method without the principle of small variations of the basic solution. The word "variational" is added to the name of the symbolic regression method that uses the principle of small variations of the basic solution.

3.4 Genetic Algorithm for Multicriterial Structural-Parametric Search of Functions

All symbolic regression methods use a genetic algorithm with a special crossover operation to find the structure and parameters of a mathematical expression. If we use the principle of small variations of the basic solution, then the crossover operation becomes a common operation of the genetic algorithm, but the operation of performing small variations is added to evaluate each possible solution.

Studies of control synthesis problems using symbolic regression methods have shown that when searching for a control function, it is important, together with the structure of the control function, to look for a vector of parameters, which is included in the control function as a part of the arguments. The introduction of parameters into the structure of a function is a natural extension of the class of the required functions.

Note that initially in genetic programming, parameters are introduced as constants included in the desired function as arguments, and their further change is carried out due to transformation using elementary functions. This approach is not entirely logical. Suppose, an acceptable structure of the control function is obtained. The only drawback of this structure is that the parameters included in this structure are not optimal. The algorithm will continue searching for the structure in order to change the parameters, i.e. the algorithm will complicate the structure and change the values of the parameters by replacing the parameters with various functions of these parameters. For example, if a parameter has a value 5, and the optimal value of this parameter for a given structure of the function is 1.57, then it is difficult to imagine how many nonlinear transformations of the parameter should be done, i.e. calculate from it the values of various functions like sin, arctan, exp, and others, in order to get the value of 1.57 from the value of 5. In addition, all these nonlinear transformations must then be included in the implemented control function.

It makes more sense to look for the values of the parameters together with the structure of the function they are included in. The search for the structure and parameters should be carried out within the framework of the same genetic algorithm, but on different data structures.

A feature of control problems is, as a rule, the presence of several quality criteria. In control problems, there is always a control goal, which is often formulated in the form of terminal conditions, and a control quality criterion, which is formulated in the form of an integral functional. In numerical synthesis, it is necessary to take into account the accuracy of achieving the goal and its influence on the assessment of the quality criterion. In practice, it is always possible to convolve criteria with certain weights, but for a genetic algorithm such a convolution does not give much advantage. Therefore, it is also possible to look for solutions in the form of a set of Pareto optimal solutions, which can always be built on the set of possible solutions used by the genetic algorithm.

Let us describe a multicriteria genetic algorithm for structural-parametric search for a function:

1. Enter the initial data for the search algorithm.
   Vector quality criterion is

$$\mathbf{J}(S, \mathbf{q}) = [j_1(S, \mathbf{q}) \ldots j_\nu(S, \mathbf{q})]^T, \tag{3.41}$$

where S is a code of the structure of the required function in symbolic regression method and $\mathbf{q}$ is the vector of parameters.
The code of the basic solution and the value of the vector of parameters for this solution are given by

$$S_0, \mathbf{q}^0 = [\overbrace{1 \ldots 1}^{p}]^T. \tag{3.42}$$

The parameters of the algorithm are set:

H is the number of possible solutions in the population,

P is the number of generations,

R is the number of possible crossovers in one generation,

α is the minimum value of the probability of crossover,

E is the number of generations in one epoch, or the number of generations between the change of the basic solution,

p_μ is the probability of performing the mutation operation,

c is the number of bits for the integer part of the parameter,

d is the number of bits for the fractional part of the parameter,

M is the number of small variations for one possible solution,

p is the dimension of the parameter vector.

2. Generate a set of possible solutions in the form of ordered sets of vectors of small variations

$$W = \{W_1, \ldots, W_H\}, \tag{3.43}$$

$$W_i = (\mathbf{w}^{i,1}, \ldots, \mathbf{w}^{i,M}), \ i = 1, \ldots, H. \tag{3.44}$$

Introduce a set of zero variations that do not vary the basic solution

$$W_0 = (\mathbf{w}^{0,1}, \ldots, \mathbf{w}^{0,M}), \tag{3.45}$$

$$\mathbf{w}^{0,i} = [0 \ldots 0]^T, \ i = 1, \ldots, M. \tag{3.46}$$

It is algorithmically determined that the zero variation vector does not change the basic solution code

$$[0 \ldots 0]^T \circ S_0 = S_0.$$

The set of zero variation vectors is introduced in order for the basic solution to be used in crossover operations.

Generate a set of binary codes for vectors of parameters of possible solutions

$$Z = \{\mathbf{z}^1, \ldots, \mathbf{z}^H\},$$

$$\mathbf{z}^i = [z_1^i \ldots z_{p(c+d)}^i]^T, \tag{3.47}$$

$$z_j^i = \xi(2),$$

where $\xi(A)$ is a random number generator function, and each time it is called, it returns a random integer from 0 to $A - 1$.

We transform this vector to the Gray code

$$\mathbf{z}^0 = \text{Gray}(\mathbf{q}^0),$$

where $\text{Gray}(\mathbf{q})$ is the function of converting a real vector into a Gray code.

3. Calculate the value of the functionals for all possible solutions, including the basic solution

$$S_i = \mathbf{w}^{i,M} \circ \ldots \circ \mathbf{w}^{i,1} \circ S_0,$$

$$\mathbf{q}^i = \text{Gray}^{-1}(\mathbf{z}^i), \tag{3.48}$$

$$\mathbf{f}^i = \mathbf{J}(S_i, \mathbf{q}^i), \ i = 0, \ldots, H,$$

where $\text{Gray}^{-1}(\mathbf{z})$ is a backward translation function from Gray code to real vector.

Calculate the values of the Pareto ranks for all values of the functionals

$$L_i = 0,$$

$$L_i \leftarrow L_i + 1, \text{ if } \mathbf{f}^j < \mathbf{f}^i, \ j = 0, \ldots, H \tag{3.49}$$

$$i = 0, \ldots, H,$$

where

$$\mathbf{f}^j < \mathbf{f}^i, \text{ if } f_k^j \leq f_k^i, k = 1, \ldots, v \text{ and } \exists r, \ 1 \leq r \leq v \Rightarrow f_r^j < f_r^i. \tag{3.50}$$

The Pareto set is defined by elements with zero Pareto rank values

$$\text{Pareto} = \{i_1, \ldots, i_K\}, \ L_{i_j} = 0, \ j = 1, \ldots, K. \tag{3.51}$$

4. Set the counter of generations $j_p = 0$.
   Step1. Start the generation cycle.
   Set the counter of the number of possible crossovers $j_c = 0$.
   Step2. Start the crossover cycle
   Select possible solutions for crossover

$$a = \xi(H), \ b = \xi(H). \tag{3.52}$$

Calculate the probability of crossover

$$Pr = \max \left\{ \frac{1 + \alpha L_a}{1 + L_a}, \frac{1 + \alpha L_b}{1 + L_b} \right\}. \tag{3.53}$$

Step3. Check the fulfillment of conditions

$$\xi \leq Pr, \tag{3.54}$$

where ξ is a random number generator function; it returns a random number from $(0; 1)$ interval.

If the condition (3.54) is not met, then go to Step 9.

Note that according to (3.53) the probability of crossover is equal to 1 if one of the parents belongs to the Pareto set and has a Pareto rank 0.

Step4. Carry out the crossover. Randomly determine the crossover points for the structural and parametric parts

$$r = \xi(M), \ s = \xi(p(c+d)). \tag{3.55}$$

Crossover is performed. Four new possible solutions are obtained. For two new possible solutions, both the structural and parametric parts are crossed; for the other two new possible solutions, only the parametric parts are crossed.

$$
\begin{aligned}
W_{H+1} &= (\mathbf{w}^{a,1}, \ldots, \mathbf{w}^{a,r-1}, \mathbf{w}^{b,r}, \ldots, \mathbf{w}^{b,M}), \\
W_{H+2} &= (\mathbf{w}^{b,1}, \ldots, \mathbf{w}^{b,r-1}, \mathbf{w}^{a,r}, \ldots, \mathbf{w}^{a,M}), \\
W_{H+3} &= W_a, \\
W_{H+4} &= W_b, \\
\mathbf{z}^{H+1} &= [z_1^a \ldots z_{s-1}^a \ z_s^b \ldots z_{p(c+d)}^b]^T, \\
\mathbf{z}^{H+2} &= [z_1^b \ldots z_{s-1}^b \ z_s^a \ldots z_{p(c+d)}^a]^T, \\
\mathbf{z}^{H+3} &= [z_1^a \ldots z_{s-1}^a \ z_s^b \ldots z_{p(c+d)}^b]^T, \\
\mathbf{z}^{H+4} &= [z_1^b \ldots z_{s-1}^b \ z_s^a \ldots z_{p(c+d)}^a]^T.
\end{aligned}
\tag{3.56}
$$

Step5. Perform the mutation for each new possible solution $H+i, i = 1, \ldots, 4$. Check the conditions for performing a mutation

$$\xi \le p_\mu. \tag{3.57}$$

If the condition (3.57) is met, then the mutation operation is performed. Randomly define positions in the structural and parametric parts

$$m_i = \xi(M), l_i = \xi(p(c+d)),$$

and generate a new variation vector and an element in the parametric part

$$\mathbf{w}^{H+i,m_i}, \ z_{l_i}^{H+i} = \xi(2). \tag{3.58}$$

Step6. Calculate the values of the functionals for each new possible solution

$$\mathbf{f}^{H+i} = \mathbf{J}(S_{H+i}, \mathbf{q}^{H+i}), \tag{3.59}$$

where

$$
\begin{aligned}
S_{H+i} &= \mathbf{w}^{H+i,M} \circ \ldots \circ \mathbf{w}^{H+i,1} \circ S_0, \\
\mathbf{q}^{H+i} &= \mathrm{Gray}^{-1}(\mathbf{z}^{H+i}), \ i = 1, \ldots, 4.
\end{aligned}
\tag{3.60}
$$

Step7. Calculate the value of the Pareto rank for the new possible solutions L_{H+i}. Determine the possible solution with the highest value of the Pareto rank

$$L_{j+} = \max\{L_k : k = 1, \ldots, H\}. \tag{3.61}$$

Step8. Check the condition for replacing solutions

$$L_{H+i} < L_{j+}.$$ (3.62)

If the condition (3.62) is met, i.e. a new possible solution $H + i$ has a value of the Pareto rank less than the largest value of the Pareto rank for the entire set of possible solutions, then we replace the solution with the largest value with the new obtained possible solution

$$W_{j+} \leftarrow W_{H+1},$$

$$\mathbf{z}^{j+} \leftarrow \mathbf{z}^{H+1},$$

$$L_{j+} \leftarrow L_{H+i},$$

$$\mathbf{f}^{j+} \leftarrow \mathbf{f}^{H+i}.$$ (3.63)

Repeat steps 7 and 8 for all new possible solutions $i = 1, \ldots, 4$.
Step9. Increase the counter of the number of possible crossovers

$$j_c \leftarrow j_c + 1.$$

Step10. Check the conditions for exiting the crossover cycle. If $j_c < R$, then go to step 2.
Step11. Increase the counter of the number of generations

$$j_p \leftarrow j_p + 1.$$

Step12. Check condition for the end of the epoch

$$j_p \mod E = 0.$$ (3.64)

If the condition (3.64) is satisfied, then replace the basic solution; otherwise go to step 15.
Step13. Change the basic solution.
For all possible solutions, including the basic one, the values of the functionals are normalized

$$\tilde{f}_i^j = \frac{f_i^j - f_i^{j-}}{f_i^{j+} - f_i^{j-}}, \quad i = 1, \ldots, \nu, \ j = 0, \ldots, H,$$ (3.65)

where

$$f_i^{j-} = \min\{f_i^j; j = 0, \ldots, H\}, \ 1 \le i \le \nu,$$ (3.66)

$$f_i^{j+} = \max\{f_i^j; j = 0, \ldots, H\}, \ 1 \le i \le \nu.$$ (3.67)

Calculate the norms of functionals for all possible solutions

$$\tilde{f}_j = \sqrt{\sum_{i=1}^{v}(\tilde{f}_i^j)^2}, \ j = 0,\ldots,H. \tag{3.68}$$

Define a new basic solution that has the smallest value of the norm of the functional

$$\tilde{f}_{j0} = \min\{\tilde{f}_j : j = 0,\ldots,H\}. \tag{3.69}$$

Change the basic solution.

$$\begin{aligned} S_{j0} &= \mathbf{w}^{j0,M} \circ \ldots \circ \mathbf{w}^{j0,1} \circ S_0, \\ S_0 &\leftarrow S_{j0}, \\ \mathbf{z}^0 &\leftarrow \mathbf{z}^{j0}, \\ \mathbf{f}^0 &\leftarrow \mathbf{f}^{j0}. \end{aligned} \tag{3.70}$$

Step14. Calculate the values of the functionals for all possible solutions using (3.48) and the value of the Pareto ranks using (3.49).
Step15. Check the condition for the end of calculations

$$j_p \geq P. \tag{3.71}$$

If the condition (3.71) is not met, then go to step 1; otherwise stop the calculations.

The solution to the multicriterial problem is the set of Pareto optimal solutions, which is determined by zero values of the Pareto rank. Then, the researcher should choose one of the possible solutions on the Pareto set as a solution to the problem. If the researcher is not satisfied with any of the solutions on the Pareto set, then he chooses one solution, which he defines as basic and starts the algorithm from the beginning.

3.5 Space of Machine-Made Functions

A machine search for a function has the peculiarity that in the process of searching, functions can take infinity values, which will stop the search process in the form of an overflow error message. To avoid the occurrence of this event, it is necessary to change the process of calculating elementary functions from the basic set so that they do not accept invalid values. Moreover, it is necessary to assume that found control functions will be realized on a board processor of robot or any other automatic device. This means that the found control functions should be realizable by computer and their modulo value never reach infinity.

Therefore, some space different from $\mathbb{R}^n$ should be introduced to eliminate overflow errors.

Let us consider such space of functions.

This space is a subspace of the real vector space

$$\mathbb{R}_{\#}^n \subseteq \mathbb{R}^n, \tag{3.72}$$

where $\mathbb{R}_{\#}^n$ is a machine-made space.

This space $\mathbb{R}_{\#}^n$ possesses the following properties.

For any vector $\mathbf{x} = [x_1 \ldots x_n]^T \in \mathbb{R}_{\#}^n$ of dimension n the following conditions are satisfied:

(1)

$$|x_i| \leq B^+ < \infty, \; i = 1, \ldots, n. \tag{3.73}$$

(2) There exists a small positive value $\delta^- > 0$ that

$$\text{if } |x_i| < \delta^-, \text{ then } x_i = 0, \; i = 1, \ldots, n. \tag{3.74}$$

(3)

$$\text{if } \mathbf{x}(t) \in \mathbb{R}_{\#}^n, \text{ then } \dot{\mathbf{x}}(t) \in \mathbb{R}_{\#}^n. \tag{3.75}$$

(4) There exists a value satisfactory accuracy $\tilde{\Delta} > \delta^-$ that

$$\text{if } |\alpha| < \tilde{\Delta}, \text{ then } x_i \pm \alpha = x_i, \; i = 1, \ldots, n. \tag{3.76}$$

Usually in the problems with differential equations, the value satisfactory accuracy is a half-step of integration.

The derivative of a function in the machine-made space $\mathbb{R}_{\#}$ is calculated by the relation

$$\frac{\partial f(z)}{\partial z} = \frac{f(z + \delta^-) - f(z)}{\delta^-}. \tag{3.77}$$

Consider an example.

$$\frac{\partial \sin(z)}{\partial z} = \frac{\sin(z + \delta^-) - \sin(z)}{\delta^-} =$$

$$\frac{\sin(z)\cos(\delta^-) + \sin(\delta^-)\cos(z) - sin(z)}{\delta^-} =$$

$$\frac{\sin(z) + \delta^- \cos(z) - \sin(z)}{\delta^-} = \cos(z). \tag{3.78}$$

Here the following equation is used

$$\cos(\delta^-) = 1 - 0.5(\delta^-)^2 = 1, \tag{3.79}$$

according to equation (3.76).

In the introduced space $\mathbb{R}_{\#}$, the machine-made functions are recorded as usual functions from mathematical analysis, but with a condition that their values are never equal to infinity.

For example,

$$z^{-1} = \begin{cases} 1/z, \text{ if } |z| > \delta^- \\ \mathrm{sgn}(z)B^+, \text{otherwise} \end{cases}.$$

The description of the most often used machine-made functions is presented in the Appendix of this chapter as a free Pascal programming code.

If it is necessary to emphasize in notation that this is a machine-made function, then the special subscript can be used. For example, $\sin_\#(z)$, $\exp_\#(z)$, etc.

Theorem 3.1. *Any machine-made function can be presented in the form of Taylor's series with a finite number of members.*

Proof. Assume $f(z)$ is a machine-made function, then for a point $z = a$ Taylor's series has the following form:

$$\sum_{k=0}^{L} \frac{f^{(k)}(a)}{k!}(z-a)^k = f(a) + f'(a)(z-a) + \frac{f''(a)}{2!}(z-a)^2 + \dots$$

$$\dots + \frac{f^{(L)}(a)}{L!}(z-a)^L.$$

The value of derivative is limited $|f^{(k)}(a)| \le B^+$ and the value of a denominator increases $k!$. For some member of Taylor's series the following inequality will be implemented

$$\frac{B^+}{k!} < \delta^-.$$

According to property (3.74) all subsequent members of the series will be zero. □

Thus, machine learning methods can also be used to search for mathematical expressions in the space of machine-made functions.

Appendix

Here is the description of the most often used machine-made functions presented in the form of free Pascal programming code, where notation Ro_N is used for functions with one argument, Xi_N is used for functions with two arguments, and Nu_N is used for functions with three arguments, where N is the number of functions.

Unit Machine-Made Functions

```
const
        infinity=1e8;
        eps=1e-8;

//*****************************
```

```
Function Ro_1 (z: real): real;
Begin
      result:=z;
End;
//******************************
Function Ro_2(z:real):real;
Begin
   if abs(z)>sqrt(infinity)
      then result:=infinity
   else result:=sqr(z);
End;
//******************************
Function Ro_3(z:real):real;
Begin
   result:=-z;
End;
//******************************
Function Ro_4(z:real):real;
Begin
   result:=Ro_10(z)*sqrt(abs(z));
End;
//******************************
Function Ro_5(z:real):real;
Begin
   if abs(z)>eps
      then result:=1/z
   else result:=Ro_10(z)/eps;
End;
//******************************
Function Ro_6(z:real):real;
Begin
   if z>-ln(eps)
      then result:=-ln(eps)
   else result:=exp(z);
End;
//******************************
Function Ro_7(z:real):real;
Begin
   if abs(z)<exp(-pokmax)
      then result:=ln(eps)
   else result:=ln(abs(z));
End;
//******************************
```

```
Function Ro_8(z:real):real;
Begin
    if abs(z)>-ln(eps)
        then result:=Ro_10(z)
    else result:=(1-exp(-z))/(1+exp(-z));
End;
//*****************************
Function Ro_9(z:real):real;
Begin
    if z>=0
        then result:=1
    else result:=0;
End;
//*****************************
Function Ro_10(z:real):real;
Begin
    if z>=0
        then result:=1
    else result:=-1;
End;
//*****************************
Function Ro_11(z:real):real;
Begin
    result:=cos(z);
End;
//*****************************
Function Ro_12(z:real):real;
Begin
    result:=sin(z);
End;
//*****************************
Function Ro_13(z:real):real;
Begin
    if abs(z)<eps
        then result:=Ro_10(z)*pi/2
    else result:=arctan(z);
End;
//*****************************
Function Ro_14(z:real):real;
Begin
    if abs(z)>Ro_15(infinity)
        then result:=Ro_10(z)*infinity
    else result:=sqr(z)*z;
End;
//*****************************
```

```
Function Ro_15(z:real):real;
Begin
   if abs(z)<eps
      then result:=Ro_10(z)*eps
   else result:=Ro_10(z)*exp(ln(abs(z))/3);
End;
//*****************************
Function Ro_16(z:real):real;
Begin
   if abs(z)<1
      then result:=z
   else result:=Ro_10(z);
End;
//*****************************
Function Ro_17(z:real):real;
Begin
   result:=Ro_10(z)*ln(abs(z)+1);
End;
//*****************************
Function Ro_18(z:real):real;
Begin
   if abs(z)>-ln(eps)
      then result:=Ro_10(z)*infinity
   else result:=Ro_10(z)*(exp(abs(z))-1);
End;
//*****************************
Function Ro_19(z:real):real;
Begin
   if abs(z)>1/eps
      then result:=Ro_10(z)*eps
   else result:=Ro_10(z)*exp(-abs(z));
End;
//*****************************
Function Ro_20(z:real):real;
Begin
   Result:=z/2;
End;
//*****************************
Function Ro_21(z:real):real;
Begin
   Result:=2*z;
End;
//*****************************
```

```
Function Ro_22(z:real):real;
Begin
   if z<0
      then Result:=exp(z)-1
   else Result:=1-exp(-abs(z));
End;
//*****************************
Function Ro_23(z:real):real;
Begin
   if abs(z)>1/eps
      then result:=-Ro_10(z)/eps
   else result:=z-z*sqr(z);
End;
//*****************************
Function Ro_24(z:real):real;
Begin
   if z>infinity
      then result:=1
   else
      if exp(-z)>infinity
         then result:=0
      else result:=1/(1+exp(-z));
End;
//*****************************
Function Ro_25(z:real):real;
Begin
   if z>0
      then result:=1
   else result:=0;
End;
//*****************************
Function Ro_26(z:real):real;
Begin
   if abs(z)<eps1
      then result:=0
   else result:=Ro_10(z);
End;
//*****************************
Function Ro_27(z:real):real;
Begin
   if abs(z)>1
      then result:=Ro_10(z)
   else result:=Ro_10(z)*(1-sqrt(1-sqr(z)));
End;
//*****************************
```

```
Function Ro_28(z:real):real;
Begin
   if z*z> ln(infinity)
      then result:=z*(1-eps)
   else result:=z*(1-exp(-sqr(z)));
End;
//*****************************
Function Xi_1(z1,z2:real):real;
Begin
   result:=z1+z2;
End;
//*****************************
Function Xi_2(z1,z2:real):real;
Begin
   if abs(z1*z2)> infinity
      then result:=Ro_10(z1*z2)*infinity
   else result:=z1*z2;
End;
//*****************************
Function Xi_3(z1,z2:real):real;
Begin
   if z1>=z2
      then result:=z1
   else result:=z2;
End;
//*****************************
Function Xi_4(z1,z2:real):real;
Begin
   if z1<z2
      then result:=z1
   else result:=z2;
End;
//*****************************
Function Xi_5(z1,z2:real):real;
Begin
   result:=z1+z2-z1*z2;
End;
//*****************************
Function Xi_6(z1,z2:real):real;
Begin
   result:=Ro_10(z1+z2)*sqrt(sqr(z1)+sqr(z2));
End;
//*****************************
Function Xi_7(z1,z2:real):real;
Begin
```

```
    result:=Ro_10(z1+z2)*(abs(z1)+abs(z2));
End;
//*****************************
Function Xi_8(z1,z2:real):real;
Begin
    result:=Ro_10(z1+z2)*Xi_2(abs(z1),abs(z2));
End;
//*****************************
Function Nu_1(z1,z2,z3:real):real;
Begin
    if z1>0
       then result:=z2
       else result:=z3;
End;
//*****************************
Function Nu_2(z1,z2,z3:real):real;
Begin
    if z1>z2
       then result:=z3
       else result:=-z3;
End;
//*****************************
Function Nu_3(z1,z2,z3:real):real;
Begin
    if z1>0
       then result:=z2+z3
       else result:=z2-z3;
End;
//*****************************
Function Nu_4(z1,z2,z3:real):real;
Begin
    if z1>z2
       then
           if z1>z3
              then result:=z1
           else result:=z3
       else
          if z2>z3
              then result:=z2
          else result:=z3;
End.
```

References

1. Cybenko, G.V.: Approximation by Superpositions of sigmoidal function, Mathematics of Control, Signals, and Systems, Vol. 5, No. 4, 1989, pp. 303–314. https://dx.doi.org/10.1007/BF02551274
2. Mitchell, T.: Machine Learning. McGraw-Hill Science/Engineering/Math. McGraw-Hill, New York (1997)
3. Han, J., Morag, C.: The influence of the sigmoid function parameters on the speed of back-propagation learning. In: Mira, J., Sandoval, F. (eds.) From Natural to Artificial Neural Computation. Lecture Notes in Computer Science, pp. 195–201. Springer, Berlin (1995)
4. Nair, V., Hinton, G.E.: Rectified linear units improve restricted Boltzmann machines. In: 27th International Conference on Machine Learning, pp. 807–814. Omnipress (2010)
5. Cruse, H.: Neural Networks as Cybernetic Systems. Brains, Minds & Media, Bielefeld, Germany (2006)
6. Hopfield, J.J., Tank, D.W.: Neural computation of decisions in optimization problems. Biolog. Cybern. **55**, 141–146 (1985)
7. Kolmogorov, A.: On the representation of continuous functions of several variables by super-positions of continuous functions of a smaller number of variables. Am. Math. Soc. Transl. **17**, 369–373 (1961)
8. Arnold, V.: On functions of three variables. Am. Math. Soc. Transl. **28**, 51–54 (1963)
9. Hecht-Nielsen, R.: Kolmogorov's mapping neural network existence theorem. In: Proceedings of the IEEE First International Conference on Neural Networks, San Diego, vol. III, pp. 11–13. IEEE, Piscataway (1987)
10. Holland, J.: Adaptation in Natural and Artificial Systems. MIT Press, Cambridge, MA (1992)
11. Mitchell, M.: An Introduction to Genetic Algorithms. MIT Press, Cambridge, MA (1996)
12. Goldberg, D.: Genetic Algorithms in Search, Optimization and Machine Learning. Addison-Wesley Professional, Reading (1989)
13. Vose, M.: The Simple Genetic Algorithm: Foundations and Theory. MIT Press, Cambridge, MA (1999)
14. Kumar, M., Husian, M., Upreti, N., Gupta, D.: Genetic Algorithm: Review and Application (2010)
15. Federal Standard 1037C (1996). https://www.its.bldrdoc.gov/fs-1037/fs-1037c.htm
16. Diveev, A.I.: Small variations of basic solution method for non-numerical optimization. IFAC-PapersOnLine **48**(25), 028–033 (2015)

Chapter 4
Symbolic Regression Methods

Abstract This chapter provides a detailed description of different symbolic regression methods. Some methods differ directly in the form of coding, as well as variational methods are based on the principle of small variations of the basic solution. By analogy with deep learning, the technology of the multilayer symbolic regression method is presented. We deliberately did not include detailed historical references in the description of the methods, focusing only on practically significant entities. The description of each method includes the encoding procedures with examples and the main features of the searching algorithm for finding the optimal solution in the code space with an emphasis on the implementation of the crossover operation of the genetic algorithm, which differs depending on the type of encoding. We do not pretend to provide a comprehensive overview of symbolic regression methods, but present only those symbolic regression methods that have already been applied in machine learning control problems, or we managed to apply them to the class of machine learning problems under consideration. As new symbolic regression methods appear for solving machine learning control problems, we will be happy to supplement the presented description.

4.1 Genetic Programming

The genetic programming method is the first and most famous symbolic regression method. It was created in the 1990s of the twentieth century.

Genetic programming uses a genetic algorithm that performs crossover and mutation operations on character strings [1, 2]. The notation of the mathematical expression in genetic programming is a string of characters without parentheses. Instead of symbols, when searching for a mathematical expression, it is advisable to use two-component numeric vectors. The first component of the vector indicates the number of arguments of the elementary function being encoded, and the second component indicates the function number. These two numbers are needed to

© The Editor(s) (if applicable) and The Author(s), under exclusive license to
Springer Nature Switzerland AG 2021
A. Diveev, E. Shmalko, *Machine Learning Control by Symbolic Regression*,
https://doi.org/10.1007/978-3-030-83213-1_4

calculate a mathematical expression using genetic programming code and to find a subexpression that is used in crossover. Each character corresponds to some operation. Each function is characterized by a certain number of arguments. Functions with no arguments are variables or parameters. The order of characters in the string determines the correspondence between functions and their arguments. In genetic programming, prefix notation of symbols is more often used. In this entry, the function symbol appears in the line before or to the left of the argument symbol.

Consider the coding procedure of a mathematical expression by GP.

A set of ordered sets of functions with a certain number of arguments is introduced

$$F = \{F_0, F_1, \ldots, F_n\}, \tag{4.1}$$

where

$$F_i = \{f_{i,1}(z_1, \ldots, z_i), \ldots, f_{i,n_i}(z_1, \ldots, z_i)\}, \tag{4.2}$$

$f_{i,j}(z_1, \ldots, z_i)$ is the function number j with the number of arguments i, $j = 1, \ldots, n_i$, $i = 1, \ldots, n$. When $i = 0$, the set F_0 is a set of arguments of the mathematical expression. In this case, the arguments of the mathematical expression are considered as elementary functions without arguments.

The function code is an integer vector of two components

$$s = [s_1 \ s_2]^T, \tag{4.3}$$

where s_1 is the number of arguments, s_2 is a number of the function in the set F_{s_1},

$$f_{i,j}(z_1, \ldots, z_i) \Leftrightarrow [i \ j]^T. \tag{4.4}$$

A mathematical expression is an ordered set of function codes

$$S = (s^1, \ldots, s^K), \tag{4.5}$$

where $s^i = [s_1^i \ s_2^i]^T$, $i = 1, \ldots, K$.

Consider an Example

Let us have a mathematical expression

$$y_1 = \exp(-ax_1)\cos(bx_2 + c). \tag{4.6}$$

Define a set of functions for this example.

$$\begin{aligned}
F_0 &= \{f_{0,1} = x_1, f_{0,2} = x_2, f_{0,3} = a, f_{0,4} = b, f_{0,5} = c\}, \\
F_1 &= \{f_{1,1}(z) = -z, f_{1,2}(z) = \exp(z), f_{1,3}(z) = \cos(z)\}, \\
F_2 &= \{f_{2,1}(z_1, z_2) = z_1 + z_2, f_{2,2}(z_1, z_2) = z_1 z_2\}.
\end{aligned} \tag{4.7}$$

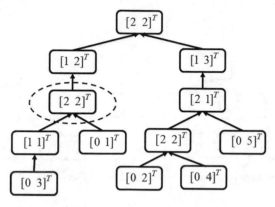

Fig. 4.1 GP computational tree for the mathematical expression (4.6)

Rewrite sets of functions in codes

$$F_0 = \left(\begin{bmatrix} 0 \\ 1 \end{bmatrix}, \begin{bmatrix} 0 \\ 2 \end{bmatrix}, \begin{bmatrix} 0 \\ 3 \end{bmatrix}, \begin{bmatrix} 0 \\ 4 \end{bmatrix}, \begin{bmatrix} 0 \\ 5 \end{bmatrix} \right),$$

$$F_1 = \left(\begin{bmatrix} 1 \\ 1 \end{bmatrix}, \begin{bmatrix} 1 \\ 2 \end{bmatrix}, \begin{bmatrix} 1 \\ 3 \end{bmatrix} \right), \tag{4.8}$$

$$F_2 = \left(\begin{bmatrix} 2 \\ 1 \end{bmatrix}, \begin{bmatrix} 2 \\ 2 \end{bmatrix} \right).$$

Encode now the mathematical expression (4.6)

$$
\begin{aligned}
y_1 &= \exp(-ax_1)\cos(bx_2+c) = \\
&f_{2,2}(\exp(-ax_1), \cos(bx_2+c)) = \\
&f_{2,2}(\exp(-ax_1), \cos(f_{2,1}(bx_2,c))) = \\
&f_{2,2}(f_{1,2}(-ax_1), f_{1,3}(f_{2,1}(bx_2,c))) = \\
&f_{2,2}(f_{1,2}(f_{2,2}(-a,x_1)), f_{1,3}(f_{2,1}(f_{2,2}(b,x_2),c))) = \\
&f_{2,2}(f_{1,2}(f_{2,2}(f_{1,1}(f_{0,3}),f_{0,1})), f_{1,3}(f_{2,1}(f_{2,2}(f_{0,4},f_{0,2}),f_{0,5}))) = \\
&f_{2,2} \circ f_{1,2} \circ f_{2,2} \circ f_{1,1} \circ f_{0,3} \circ f_{0,1} \circ f_{1,3} \circ f_{2,1} \circ f_{2,2} \circ f_{0,4} \circ f_{0,2} \circ f_{0,5}.
\end{aligned} \tag{4.9}
$$

Rewrite this record in codes

$$
\begin{aligned}
S_1 = \Bigg(&\begin{bmatrix} 2 \\ 2 \end{bmatrix}, \begin{bmatrix} 1 \\ 2 \end{bmatrix}, \begin{bmatrix} 2 \\ 2 \end{bmatrix}, \begin{bmatrix} 1 \\ 1 \end{bmatrix}, \begin{bmatrix} 0 \\ 3 \end{bmatrix}, \begin{bmatrix} 0 \\ 1 \end{bmatrix}, \begin{bmatrix} 1 \\ 3 \end{bmatrix}, \\
&\begin{bmatrix} 2 \\ 1 \end{bmatrix}, \begin{bmatrix} 2 \\ 2 \end{bmatrix}, \begin{bmatrix} 0 \\ 2 \end{bmatrix}, \begin{bmatrix} 0 \\ 4 \end{bmatrix}, \begin{bmatrix} 0 \\ 5 \end{bmatrix} \Bigg).
\end{aligned} \tag{4.10}
$$

Figure 4.1 shows a computational tree for the given mathematical expression.

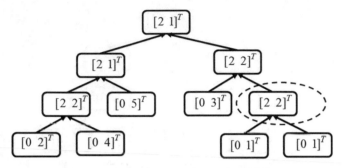

Fig. 4.2 GP computational tree for the mathematical expression (4.11)

The code of the mathematical expression (4.10) is written correctly if the condition for the correct code (3.10) is met for it. Index of the code element (3.9) is necessary for the correct execution of crossover.

Having defined the coding procedure, it is necessary to consider the performance of crossover. In genetic programming, crossover is performed by swapping subtrees. It is not difficult to define a subtree in Fig. 4.1. But defining a subtree in a sequential code notation (4.10) causes certain difficulties.

Consider the specified features of the crossover operation both for trees and for sequential code notation.

To show the crossover for GP codes, one more mathematical expression needs to be considered

$$y_2 = bx_2 + c + ax_1^2. \tag{4.11}$$

The code of this mathematical expression, using the sets of functions (4.7), has the following form:

$$S_2 = \left(\begin{bmatrix} 2 \\ 1 \end{bmatrix}, \begin{bmatrix} 2 \\ 1 \end{bmatrix}, \begin{bmatrix} 2 \\ 2 \end{bmatrix}, \begin{bmatrix} 0 \\ 2 \end{bmatrix}, \begin{bmatrix} 0 \\ 4 \end{bmatrix}, \begin{bmatrix} 0 \\ 5 \end{bmatrix}, \begin{bmatrix} 2 \\ 2 \end{bmatrix}, \begin{bmatrix} 0 \\ 3 \end{bmatrix}, \begin{bmatrix} 2 \\ 2 \end{bmatrix}, \begin{bmatrix} 0 \\ 1 \end{bmatrix}, \begin{bmatrix} 0 \\ 1 \end{bmatrix} \right). \tag{4.12}$$

The computation tree for the mathematical expression (4.11) is shown in Fig. 4.2.

Perform the crossover for the computational trees shown in Figs. 4.1 and 4.2. The outlined nodes are randomly selected nodes for crossover.

Exchange subtrees that start at these nodes and get new computational trees shown in Figs. 4.3 and 4.4.

The mathematical expressions of these new trees are following:

$$y_3 = \exp(x^2)\cos(bx_2 + c), \tag{4.13}$$

$$y_4 = x_2 b + c - a^2 x_1. \tag{4.14}$$

As seen, finding a subtree from a tree graph is not particularly difficult. But with a machine implementation of genetic programming, the computer memory uses codes

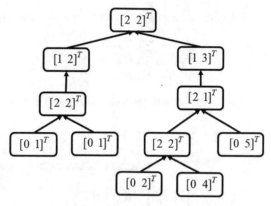

Fig. 4.3 GP computational tree for the mathematical expression (4.13)

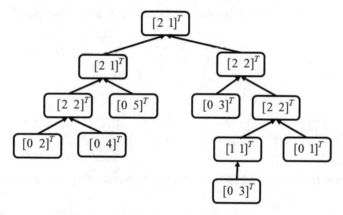

Fig. 4.4 GP computational tree for the mathematical expression (4.14)

(4.10) and (4.12). To perform crossover for these codes, it is necessary to determine the subsequences of the codes of the exchanged subtrees.

For this purpose, the code element index (3.9) is used. Since each subtree is a tree, we consider the first element of the subtree to be the first element of the tree. We calculate sequentially the indices of all the following elements until the element's index becomes equal to zero. This will point to the last code element of the subtree.

Consider the code (4.10) of the mathematical expression (4.6). The element 3 in the code is chosen as crossover point.

Consider it as the first element of the subtree and calculate the index of this element as the first element using the formula (3.9)

$$T(3) = 1 + 2 - 1 = 2.$$

Calculate sequentially the indices of the remaining code elements until the index becomes zero

$$T(4) = 2 + 1 - 1 = 2,$$
$$T(5) = 2 + 0 - 1 = 1,$$
$$T(6) = 1 + 0 - 1 = 0.$$

As a result, the sequence of the subtree code includes four elements

$$S_1 = \left(\begin{bmatrix} 2 \\ 2 \end{bmatrix}, \begin{bmatrix} 1 \\ 2 \end{bmatrix}, \underbrace{\begin{bmatrix} 2 \\ 2 \end{bmatrix}, \begin{bmatrix} 1 \\ 1 \end{bmatrix}, \begin{bmatrix} 0 \\ 3 \end{bmatrix}, \begin{bmatrix} 0 \\ 1 \end{bmatrix}}, \begin{bmatrix} 1 \\ 3 \end{bmatrix}, \right.$$

$$\left. \begin{bmatrix} 2 \\ 1 \end{bmatrix}, \begin{bmatrix} 2 \\ 2 \end{bmatrix}, \begin{bmatrix} 0 \\ 2 \end{bmatrix}, \begin{bmatrix} 0 \\ 4 \end{bmatrix}, \begin{bmatrix} 0 \\ 5 \end{bmatrix} \right). \tag{4.15}$$

Perform the same actions for the code of the second mathematical expression (4.11).

$$T(9) = 2 + 1 - 1 = 2,$$
$$T(10) = 2 + 0 - 1 = 1,$$
$$T(11) = 1 + 0 - 1 = 0.$$

$$S_2 = \left(\begin{bmatrix} 2 \\ 1 \end{bmatrix}, \begin{bmatrix} 2 \\ 1 \end{bmatrix}, \begin{bmatrix} 2 \\ 2 \end{bmatrix}, \begin{bmatrix} 0 \\ 2 \end{bmatrix}, \begin{bmatrix} 0 \\ 4 \end{bmatrix}, \begin{bmatrix} 0 \\ 5 \end{bmatrix}, \begin{bmatrix} 2 \\ 2 \end{bmatrix}, \begin{bmatrix} 0 \\ 3 \end{bmatrix}, \underbrace{\begin{bmatrix} 2 \\ 2 \end{bmatrix}, \begin{bmatrix} 0 \\ 1 \end{bmatrix}, \begin{bmatrix} 0 \\ 1 \end{bmatrix}} \right). \tag{4.16}$$

Exchange now the found subsequences and obtain codes for new mathematical expressions

$$S_3 = \left(\begin{bmatrix} 2 \\ 2 \end{bmatrix}, \begin{bmatrix} 1 \\ 2 \end{bmatrix}, \underbrace{\begin{bmatrix} 2 \\ 2 \end{bmatrix}, \begin{bmatrix} 0 \\ 1 \end{bmatrix}, \begin{bmatrix} 0 \\ 1 \end{bmatrix}}, \begin{bmatrix} 1 \\ 3 \end{bmatrix}, \begin{bmatrix} 2 \\ 1 \end{bmatrix}, \begin{bmatrix} 2 \\ 2 \end{bmatrix}, \begin{bmatrix} 0 \\ 2 \end{bmatrix}, \begin{bmatrix} 0 \\ 4 \end{bmatrix}, \begin{bmatrix} 0 \\ 5 \end{bmatrix} \right), \tag{4.17}$$

$$S_4 = \left(\begin{bmatrix} 2 \\ 1 \end{bmatrix}, \begin{bmatrix} 2 \\ 1 \end{bmatrix}, \begin{bmatrix} 2 \\ 2 \end{bmatrix}, \begin{bmatrix} 0 \\ 2 \end{bmatrix}, \begin{bmatrix} 0 \\ 4 \end{bmatrix}, \begin{bmatrix} 0 \\ 5 \end{bmatrix}, \begin{bmatrix} 2 \\ 2 \end{bmatrix}, \right.$$

$$\left. \begin{bmatrix} 0 \\ 3 \end{bmatrix}, \underbrace{\begin{bmatrix} 2 \\ 2 \end{bmatrix}, \begin{bmatrix} 1 \\ 1 \end{bmatrix}, \begin{bmatrix} 0 \\ 3 \end{bmatrix}, \begin{bmatrix} 0 \\ 1 \end{bmatrix}} \right). \tag{4.18}$$

The new codes obtained after crossover correspond to the following mathematical expressions (4.13), (4.14).

Other operations of the genetic algorithm in GP are performed in a standard manner and do not present programming difficulties.

The main drawback of the genetic programming consists in different code lengths for different mathematical expressions, and, accordingly, a change in the code lengths after the crossover operation.

4.2 Grammatical Evolution

The method of grammatical evolution [3,4] is designed to search for mathematical expressions in the program code of Backus–Naur form. To describe the language grammar in Backus–Naur form, four finite sets of characters are used:

T—set of terminal symbols;
N—set of nonterminal symbols;
P—set of production rules;
S—set of starting symbols, $S \subset N$.

As an example, consider the following sets:

$$
\begin{aligned}
T &= \{\sin, \cos, \tan, +, -, /, \times, x, z\}, \\
S &= \{< \text{expr} >\}, \\
N &= \{< \text{expr} >, < \text{op} >, < \text{pre\_op} >, < \text{var} >\}.
\end{aligned}
\tag{4.19}
$$

The set of production rules P is written as

$$
\begin{aligned}
(1) < \text{expr} > &::= < \text{expr} >< \text{op} >< \text{expr} > \quad (0) \\
&| < \text{expr} >< \text{op} >< \text{expr} > \quad (1) \\
&| < \text{pre\_op} > (< \text{expr} >) \quad (2) \\
&| < \text{var} > \quad (3)
\end{aligned}
$$

$$
\begin{aligned}
(2) < \text{op} > &::= + \quad (0) \\
&| - \quad (1) \\
&| / \quad (2) \\
&| \times \quad (3)
\end{aligned}
$$

$$
\begin{aligned}
(3) < \text{pre\_op} > &::= \sin \quad (0) \\
&| \cos \quad (1) \\
&| \tan \quad (2)
\end{aligned}
$$

$$
\begin{aligned}
(4) < \text{var} > &::= x \quad (0) \\
&| z \quad (1)
\end{aligned}
$$

To encode an expression, an ordered set of integers is used

$$
C = (c_1, \ldots, c_K).
\tag{4.20}
$$

Each number from the set (4.20) indicates the number of an element from the corresponding set of production rules.

Consider an Example

Suppose we have the following set of numbers:

$$
C = (2, 1, 1, 3, 0, 0, 3, 1).
\tag{4.21}
$$

The initial character in the expression is $y =< \text{expr} >$. The number $c_1 = 2$ in the rules for the symbols $< \text{expr} >$ defines the expression $< \text{pre\_op} > (< \text{expr} >)$.

Then

$$y =< \text{pre\_op} > (< \text{expr} >). \tag{4.22}$$

The number $c_2 = 1$ in the set of symbols $< \text{pre\_op} >$ defines the function cos. Then

$$y = \cos(< \text{expr} >). \tag{4.23}$$

Still not disclosed the symbol $< \text{expr} >$. The next number $c_3 = 1$. Then

$$y = \cos(< \text{expr} >< \text{op} >< \text{expr} >). \tag{4.24}$$

Continue step by step.

$c_4 = 3$:

$$y = \cos(< \text{var} >< \text{op} >< \text{expr} >). \tag{4.25}$$

$c_5 = 0$:

$$y = \cos(x < \text{op} >< \text{expr} >). \tag{4.26}$$

$c_6 = 0$:

$$y = \cos(x+ < \text{expr} >). \tag{4.27}$$

$c_7 = 3$:

$$y = \cos(x+ < \text{var} >). \tag{4.28}$$

$c_8 = 1$:

$$y = \cos(x+z). \tag{4.29}$$

To encode integers, eight-bit strings, codons, are used. Each codon encodes a number from 0 to 255. To determine the number of an element in the set of rules, the remainder of the division of the number by the number of elements in the set is calculated. For example, the codon for some number is 00100111. This corresponds to the number $c_7 = 39$ and $(39 \mod 4) = 3$.

To search for an expression, fixed-length codon sets are used. If there are not enough numbers to decode the expression, the last numbers are replaced with terminal element codes. If numbers are redundant, the last numbers of codes are not taken into account.

The expression is searched using the genetic algorithm. The crossover is performed by exchanging parts of binary codes after a randomly selected crossover point.

Consider the possibility of applying grammatical evolution to machine learning control.

Let us introduce a set of ordered sets of functions with a certain number of arguments

$$F = \{F_0, F_1, \ldots, F_n\}, \tag{4.30}$$

where

$$F_i = \{f_{i,1}(z_1, \ldots, z_i), \ldots, f_{i,n_i}(z_1, \ldots, z_i)\}, \tag{4.31}$$

$f_{i,j}(z_1,\ldots,z_i)$ is the function at number j with the number of arguments i, $j = 1,\ldots,n_i$, $i = 1,\ldots,n$.

As a function code, a binary vector of $2L$ elements is used

$$\mathbf{c} = [c_1 \ldots c_L \, c_{L+1} \ldots c_{2L}]^T, \tag{4.32}$$

where $c_j \in \{0,1\}$, $j = 1,\ldots,2L$.

The first L elements in the function code (4.32) determine the number of arguments in the function or a subset (4.30). For this purpose, we translate the first L elements in the function code into decimal code and take the remainder of dividing the resulting number by the maximum possible number of arguments in the function

$$r = \sum_{k=1}^{L} 2^k c_k, \tag{4.33}$$

$$i = r \mod n. \tag{4.34}$$

To determine the function number in a certain set of functions F_i, we use the following L elements in the function code:

$$r = \sum_{k=L+1}^{2L} 2^{k-L} c_{k-L}, \tag{4.35}$$

$$i = r \mod n. \tag{4.36}$$

A mathematical expression code is an ordered set of function codes

$$C = (\mathbf{c}^1,\ldots,\mathbf{c}^K), \tag{4.37}$$

where $\mathbf{c}^i = [c_1^i \ldots c_{2L}^i]^T$, $c_j^i \in \{0,1\}$, $i = 1,\ldots,K$, $j = 1,\ldots,2L$.

Consider an Example

Let us have the following sets of functions:

$$\begin{aligned}
F_0 &= \{a,b,c,x\}, \\
F_1 &= \{-z,\exp(z),\cos = (z)\}, \\
F_2 &= \{z_1 + z_2, z_1 \cdot z_2\}.
\end{aligned} \tag{4.38}$$

So, there are n sets, the set F_0 has $m_0 = 4$ elements, the set F_1 has $m_1 = 3$ elements, the set F_2 has $m_2 = 2$ elements.

Let $L = 8$ and a binary code of $2L = 16$ elements is given.

$$\mathbf{c} = [1\,0\,1\,1\,0\,1\,1\,0\,0\,0\,1\,0\,0\,0\,0\,1]^T. \tag{4.39}$$

Convert the first $L = 8$ elements into decimal code

$$[1\,0\,1\,1\,0\,1\,1\,0]^T \Rightarrow (182)_2, \quad r = 182.$$

Define the set of functions.

$$i = r \quad \text{mod } n = 183 \quad \text{mod } 3 = 2.$$

So, this is a set of functions with two arguments F_2.
Translate the second part of the code of $L = 8$ elements into decimal code

$$[0\,0\,1\,0\,0\,0\,0\,1]^T \Rightarrow (33)_2, \quad t = 33.$$

Define the function

$$j = t \quad \text{mod } i = 33 \quad \text{mod } 2 = 1.$$

Thus, this is a multiplication function

$$f_{2,1}(z_1, z_2) = z_1 \cdot z_2.$$

Each binary code $\mathbf{c}^k$ in grammatical evolution corresponds to the vector code of two numbers $\mathbf{s}^k = [i\ j]^T$ similar to genetic programming. After converting the binary codes $\mathbf{c}^k$ into integer codes $\mathbf{s}^k = [i\ j]^T$, the calculations of the mathematical expression are performed in the same way.

A significant difference between grammatical evolution and genetic programming is the operation of reproducing new codes from existing ones. The crossover in grammatical evolution is performed according to the usual rules for exchanging the tails of codes after a randomly selected crossover point.

Select two codes of mathematical expressions for crossover

$$\mathbf{C}^\alpha = (c_1^\alpha, \ldots, c_{2L}^\alpha, c_{2L+1}^\alpha, \ldots, c_{2LK}^\alpha),$$

$$\mathbf{C}^\beta = (c_1^\beta, \ldots, c_{2L}^\beta, c_{2L+1}^\beta, \ldots, c_{2LK}^\beta), \tag{4.40}$$

where $c_i^\alpha, c_i^\beta \in \{0, 1\}$, $i = 1, \ldots, 2LK$.

Randomly determine the crossover point $v \in \{1, \ldots, 2LK\}$. New codes are obtained as a result of exchanging parts of the codes after the crossover point

$$\mathbf{C}^\gamma = (c_1^\alpha, \ldots, c_v^\alpha, c_{v+1}^\beta, \ldots, c_{2LK}^\beta),$$

$$\mathbf{C}^\delta = (c_1^\beta, \ldots, c_v^\beta, c_{v+1}^\alpha, \ldots, c_{2LK}^\alpha). \tag{4.41}$$

The crossover operation in grammatical evolution:

- Is performed at any point;
- Does not require at the crossover point matching the numbers of arguments for functions;
- Does not require finding subexpressions;
- Does not change the length of the code.

To calculate the value of a mathematical expression, it is necessary to translate the binary code of grammatical evolution into the genetic programming code

$$C = (c_1, \ldots, c_{2LK}) \Rightarrow S = (s^1, \ldots, s^K). \tag{4.42}$$

The crossover operation in grammatical evolution can lead to the violation of the code correctness condition (3.10) for the corresponding code in genetic programming. The code index of the last element can have a non-zero value $T(K) \neq 0$ or vice versa the code index of a non-last element can have a zero value $T(j) = 0$, $j < K$.

In case of violation of the code correctness condition (3.10), the code correction rules should be applied.

Let the last code element have a positive index $T(K) > 0$.

Calculate the indices of all elements of the record using the formula

$$T(j) = 1 - j + \sum_{i=1}^{j} s_1^i, \ j = 1, \ldots, K. \tag{4.43}$$

Element index shows the minimum number of elements that must be to the right of a given element.

Find the first element $s^j = [s_1^j \ s_2^j]^T$, for which

$$T(j) > K - j, \tag{4.44}$$

then a violation of the correctness of the code is recognized, since it will also lead to $T(K) > 0$. Then, according to the verification procedure, the index of the previous element satisfies the conditions for correct notation, therefore $T(j-1) \leq K - j + 1$ for it. Then

$$s_1^j = T(j) - T(j-1) + 1.$$

Replace the element code s^j with the new code $\tilde{s}^j = [\tilde{s}_1^j \ \tilde{s}_2^j]^T$, in which we decrease the value of the first component

$$\tilde{s}_1^j = s_1^j - T(j) + K - j,$$

and adjust the value of the second component

$$\tilde{s}_2^j = s_2^j \mod |F_t|, \ t = \tilde{s}_1^j.$$

Next, we check the remaining elements of the record and, if a code violation condition is detected (4.44), then the specified correction should be carried out.

If another type of code incorrectness is detected, for example, the non-last element is equal to zero $T(j) = 0$, $j < K$, then we also perform the appropriate correction, for example, we use only correct entries in the code, and exclude code entries with violation.

As can be seen, the method of grammatical evolution easily implements the crossover operation, but it has the main drawback, which is the verification and correction of the received code of the mathematical expression.

4.3 Cartesian Genetic Programming

Many researchers and programmers note some shortcomings in conventional genetic programming, such as the need for recursive computations when looking for a subexpression for crossing and different lengths of codes for different expressions, which provoke additional computational difficulties. The coding of mathematical expressions in Cartesian genetic programming [5, 6] is aimed at overcoming these difficulties.

Cartesian Genetic Programming codes a mathematical expression in the form of a set of integer vectors. Each vector contains all the necessary codes for calculations— these are the codes of the function, its arguments, and the code of the variable, where the calculation result should be written.

$$\mathbf{G} = (\mathbf{g}^1, \ldots, \mathbf{g}^M),$$
(4.45)

where

$$\mathbf{g}^i = [g_1^i \ldots g_R^i]^T,$$
(4.46)

g_1^i is the number of a function, g_j^i is the number of an argument, $j = 2, \ldots, R$, $i = 1, \ldots, M$.

To code a mathematical expression it is necessary to determine the basic set of elementary functions and the set of arguments of the mathematical expression. Let the basic set includes k_1 functions with one argument, k_2 functions with two arguments, and k_3 functions with three arguments. Then the basic set of elementary functions is

$$\mathbf{F} = (f_1(z), \ldots, f_{k_1}(z), f_{k_1+1}(z_1, z_2), \ldots$$
$$f_{k_1+k_2}(z_1, z_2), \ldots, f_{k_1+k_2+1}(z_1, z_2, z_3), \ldots$$
$$f_{k_1+k_2+k_3}(z_1, z_2, z_3)).$$
(4.47)

The set of arguments is

$$\mathbf{F}_0 = (x_1, \ldots, x_n, q_1, \ldots, q_p, e_1, \ldots, e_{k_2}),$$
(4.48)

where x_i is a variable, $i = 1, \ldots, n$, q_j is a constant parameter, $j = 1, \ldots, p$, e_k is a unit element for a function with two arguments,

$$f_{k_1+l}(z_1, e_l) = z_1, \text{and } f_{k_1+l}(e_l, z_2) = z_2.$$
(4.49)

Functions with three arguments are needed to code if-operator, which is often applied in adaptive control systems.

For example,

$$f(z_1, z_2, z_3) = \begin{cases} z_2, \text{ if } z_1 \leq 0 \\ z_3, \text{ otherwise} \end{cases}. \tag{4.50}$$

If for coding some mathematical expression we use only functions with not more than three arguments, then $R = 4$. The first component is a number of function from the set (4.47) and other components are numbers of arguments from the set (4.48) or numbers of the vectors (4.46) that have already been computed.

For example,

$$\mathbf{g}^i = [g_1^i \ g_2^i \ g_3^i \ g_4^i]^T, \tag{4.51}$$

where g_1^i is a number of function, if $g_1^i \leq k_1$, then it is a function with one argument $f_{g_1^i}(z)$, if $k_1 < g_1^i \leq k_1 + k_2$, then it is a number of function with two arguments $f_{g_1^i}(z_1, z_2)$, if $k_1 + k_2 < g_1 \leq k_1 + k_2 + k_3$, then it is a function with three arguments, $f_{g_1^i}(z_1, z_2, z_3)$, g_2^i, g_3^i, g_4^i are numbers of arguments, if $1 \leq g_k^i \leq n + p + k_2, k = 2, 3, 4,$ then it is an element from the set of arguments (4.48), if $g_k^i > n + p + k_2, k = 2, 3, 4,$ then g_k^i must be not more than $n + p + k_2 + i - 1$, in this case, an argument is the result of the calculation of the function called by the vector $\mathbf{g}^r = [g_1^r \ g_2^r \ g_3^r \ g_4^r]^T$, where $r = g_k^i - n - p - k_2, k = 2, 3, 4$. If g_1^i is a number of function with one argument, then components g_3^i and g_4^i are not used. If g_1^i is a number of function with two arguments, then the component g_4^i is not used.

Values of all components of the vector (4.51) must satisfy the restrictions

$$\begin{aligned} g_1^i &\in \{1, \ldots, k_1 + k_2 + k_3\}, \\ g_k^i &\in \{1, \ldots, n + p + k_2 + i - 1\}, k = 2, 3, 4. \end{aligned} \tag{4.52}$$

In order to calculate a mathematical expression by a code of Cartesian genetic programming (4.45), the vector of results is needed

$$\mathbf{y} = [y_1 \ldots y_M]^T, \tag{4.53}$$

where

$$y_i = \begin{cases} f_{g_1^i}(g_2^i), \text{ if } g_1^i \leq k_1, \\ f_{g_1^i}(g_2^i, g_3^i), \text{ if } k_1 < g_1^i \leq k_2, \\ f_{g_1^i}(g_2^i, g_3^i, g_4^i), \text{ if } k_2 < g_1^i \leq k_3, \end{cases} \tag{4.54}$$

where $i = 1, \ldots, M$.

Consider an example of coding the following mathematical expression:

$$y_1 = \exp(q_1 x_1)(\sin(q_2 x_2) + \cos(q_3 x_3)). \tag{4.55}$$

For this example, the basic sets are

$$\begin{aligned} \mathbf{F} = (&f_1(z) = z, f_2(z) = -z, f_3(z) = \cos(z), f_4(z) = \sin(z), \\ &f_5(z) = \exp(z), f_6(z_1, z_2) = z_1 + z_2, f_7(z_1, z_2) = z_1 z_2), \end{aligned} \tag{4.56}$$

$$\mathbf{F}_0 = (x_1, x_2, x_3, q_1, q_2, q_3). \tag{4.57}$$

To code a mathematical expression q_1x_1, find a function of multiplication in the set of functions (4.56). This function is number 7, $f_7(z_1, z_2) = z_1z_2$. Then the numbers of elements in the set of arguments (4.57) are found. The parameter q_1 is an element number 4, the variable x_1 is an element 1. The fourth component in the code is not used; therefore, it can be of any value according to restrictions (4.52), for example $g_4^1 = 2$. As a result, the code of q_1x_1 is $\mathbf{g}^1 = [7\,4\,1\,2]^T$. Then, the code of q_2x_2 is $\mathbf{g}^2 = [7\,5\,2\,3]^T$, and the code of q_3x_3 is $\mathbf{g}^3 = [7\,6\,3\,4]^T$.

Then the code of $\exp(q_1x_1)$ is written. The number of function $\exp(z)$ is 5. This function has one argument; it is a result of calculation of $\mathbf{g}^1$. The number of the argument is $|F_0| + 1 = 6 + 1 = 7$. As a result, the following vector is received $\mathbf{g}^4 = [5\,7\,5\,6]^T$. Components $g_3^4 = 5$ and $g_4^4 = 6$ are not used.

Consequently, the code of the mathematical expression (4.55) has the following form:

$$G_1 = \left(\begin{bmatrix} 7 \\ 4 \\ 1 \\ 2 \end{bmatrix}, \begin{bmatrix} 7 \\ 5 \\ 2 \\ 3 \end{bmatrix}, \begin{bmatrix} 7 \\ 6 \\ 3 \\ 4 \end{bmatrix}, \begin{bmatrix} 5 \\ 7 \\ 5 \\ 6 \end{bmatrix}, \begin{bmatrix} 4 \\ 8 \\ 1 \\ 2 \end{bmatrix}, \begin{bmatrix} 3 \\ 9 \\ 3 \\ 4 \end{bmatrix}, \begin{bmatrix} 6 \\ 11 \\ 12 \\ 5 \end{bmatrix}, \begin{bmatrix} 7 \\ 10 \\ 13 \\ 6 \end{bmatrix} \right). \tag{4.58}$$

It is not known in advance how many vectors are to be used to encode the desired function. Code length is pre-set as $L = 8$. As a rule, the length of the codes is chosen to be redundant, and the extra elements are simply recorded, but not taken into account.

Now consider an example of crossover operation.

Let the first selected parent be (4.58). And the second is following:

$$G_2 = \left(\begin{bmatrix} 7 \\ 1 \\ 4 \\ 2 \end{bmatrix}, \begin{bmatrix} 7 \\ 2 \\ 5 \\ 1 \end{bmatrix}, \begin{bmatrix} 7 \\ 3 \\ 6 \\ 2 \end{bmatrix}, \begin{bmatrix} 5 \\ 9 \\ 2 \\ 3 \end{bmatrix}, \begin{bmatrix} 7 \\ 8 \\ 10 \\ 1 \end{bmatrix}, \begin{bmatrix} 4 \\ 11 \\ 2 \\ 3 \end{bmatrix}, \begin{bmatrix} 7 \\ 12 \\ 7 \\ 3 \end{bmatrix}, \begin{bmatrix} 3 \\ 13 \\ 4 \\ 2 \end{bmatrix} \right). \tag{4.59}$$

The code of the second parent describes a mathematical expression

$$y_2 = \cos(q_1x_1 \sin(q_2x_2 \exp(q_3x_3))). \tag{4.60}$$

Assume that a crossover point is $r = 5$. Then two new codes are obtained

$$G_3 = \left(\begin{bmatrix} 7 \\ 4 \\ 1 \\ 2 \end{bmatrix}, \begin{bmatrix} 7 \\ 5 \\ 2 \\ 3 \end{bmatrix}, \begin{bmatrix} 7 \\ 6 \\ 3 \\ 4 \end{bmatrix}, \begin{bmatrix} 5 \\ 7 \\ 5 \\ 6 \end{bmatrix}, \begin{bmatrix} 4 \\ 8 \\ 1 \\ 2 \end{bmatrix}, \begin{bmatrix} 4 \\ 11 \\ 2 \\ 3 \end{bmatrix}, \begin{bmatrix} 7 \\ 12 \\ 7 \\ 3 \end{bmatrix}, \begin{bmatrix} 3 \\ 13 \\ 4 \\ 2 \end{bmatrix} \right), \tag{4.61}$$

$$G_4 = \left(\begin{bmatrix} 7 \\ 1 \\ 4 \\ 2 \end{bmatrix}, \begin{bmatrix} 7 \\ 2 \\ 5 \\ 1 \end{bmatrix}, \begin{bmatrix} 7 \\ 3 \\ 6 \\ 2 \end{bmatrix}, \begin{bmatrix} 5 \\ 9 \\ 2 \\ 3 \end{bmatrix}, \begin{bmatrix} 7 \\ 8 \\ 10 \\ 1 \end{bmatrix}, \begin{bmatrix} 3 \\ 9 \\ 3 \\ 4 \end{bmatrix}, \begin{bmatrix} 6 \\ 11 \\ 12 \\ 5 \end{bmatrix}, \begin{bmatrix} 7 \\ 10 \\ 13 \\ 6 \end{bmatrix} \right). \tag{4.62}$$

These new codes correspond to the following mathematical expressions:

$$y_3 = \cos(q_1 x_1 \sin(\sin(q_2 x_2))), \tag{4.63}$$

$$y_4 = \exp(q_3 x_3)(q_2 x_2 \exp(q_3 x_3) + \cos(\exp(q_3 x_3))). \tag{4.64}$$

4.4 Inductive Genetic Programming

Another type of genetic programming is the method of inductive genetic programming [7, 8]. Inductive genetic programming uses smooth polynomials of two variables as basic functions. A discontinuous function cannot be obtained by this method.

A mathematical expression encoded in IGP is a multidimensional polynomial. To encode a mathematical expression, the parameters, variables, and basic elementary second-order polynomials are placed in an ordered set of functions

$$\mathbf{F} = \{f_1 = q_1, \dots, f_p = q_p, f_{p+1} = x_1, \dots \\ \dots, f_{p+n} = x_n, f_{p+n+1}(z_1, z_2), \dots, f_{p+n+v}(z_1, z_2)\}, \tag{4.65}$$

where p is the number of parameters, n is the number of variables.

The code of the inductive genetic programming function is an ordered set of numbers of elements from (4.65)

$$I = (i_1, \dots, i_L), \tag{4.66}$$

where i_j is a number of function $f_{i_j} \in F$, $j = 1, \dots, L$.

The code (4.66) is the composition of functions, variables, and parameters from (4.65)

$$y = A_1 \circ A_2 \circ \dots \circ A_L, \tag{4.67}$$

where

$$A_j = \begin{cases} q_{i_j}, & \text{if } 1 \leq i_j \leq p \\ x_{i_j}, & \text{if } p \leq i_j \leq p+v \\ f_{i_j}, & \text{otherwise} \end{cases} . \tag{4.68}$$

To check the correctness of the code in inductive genetic programming, the element index (3.9) and the code correctness condition (3.10) also can be used.

Since the code uses only functions without arguments and functions with two arguments, the index of the code element (4.66) in inductive genetic programming is calculated by the formula

$$T(j) = \sum_{k=1}^{j} a_k, \tag{4.69}$$

where

$$a_k = \begin{cases} -1, & \text{if } i_k \leq p+n \\ 1, & \text{otherwise} \end{cases} . \tag{4.70}$$

The list of basic polynomials of two variables used in inductive genetic programming is the following:

$$
\begin{aligned}
f_{n+p+1}(z_1, z_2) &= z_1 + z_2 + z_1 z_2, \\
f_{n+p+2}(z_1, z_2) &= z_1 + z_2, \\
f_{n+p+3}(z_1, z_2) &= z_1 + z_1 z_2, \\
f_{n+p+4}(z_1, z_2) &= z_1 + z_1 z_2 + z_1^2, \\
f_{n+p+5}(z_1, z_2) &= z_1 + z_2^2, \\
f_{n+p+6}(z_1, z_2) &= z_1 + z_2 + z_1^2, \\
f_{n+p+7}(z_1, z_2) &= z_1 + z_1^2 + z_2^2, \\
f_{n+p+8}(z_1, z_2) &= z_1^2 + z_2^2, \\
f_{n+p+9}(z_1, z_2) &= z_1 + z_2 + z_1 z_2 + z_1^2 + z_2^2, \\
f_{n+p+10}(z_1, z_2) &= z_1 + z_2 + z_1 z_2 + z_1^2, \\
f_{n+p+11}(z_1, z_2) &= z_1 + z_1 z_2 + z_1^2 + z_2^2, \\
f_{n+p+12}(z_1, z_2) &= z_1 z_2 + z_1^2 + z_2^2, \\
f_{n+p+13}(z_1, z_2) &= z_1 + z_1 z_2 + z_1^2, \\
f_{n+p+14}(z_1, z_2) &= z_1 + z_2 + z_1^2 + z_2^2, \\
f_{n+p+15}(z_1, z_2) &= z_1 z_2, \\
f_{n+p+16}(z_1, z_2) &= z_1 z_2 + z_1^2.
\end{aligned}
\tag{4.71}
$$

Codes for different mathematical expressions in inductive genetic programming can have different lengths.

Graphically, the inductive genetic programming code looks like a computational tree. The leaves of the tree contain parameters and variables, and the nodes of the tree contain the numbers of the basic polynomials.

The crossover operation in inductive genetic programming is performed by exchanging subtrees or subsequences of codes. Crossover for two IGP codes does not require special conditions for crossover points. To perform the crossover operation, a subtree code needs to be defined. The definition of the subtree or subsequence for crossover is performed according to the correct record conditions (3.10).

For crossover, two codes are selected. In each of the codes, crossover points are randomly selected and the codes of subtrees are found. The codes of the subtrees are exchanged with each other, and the codes of new mathematical expressions are obtained.

Consider an Example

Let us have the following set of functions:

$$
F = \{q_1, q_2, q_3, x_1, x_2, x_3, f_7(z_1, z_2), \ldots, f_{22}(z_1, z_2)\}
\tag{4.72}
$$

and IGP codes of the mathematical expressions

$$
I_1 = (15, 12, 1, 19, 4, 2, 14, 12, 5, 20, 6, 4, 3),
\tag{4.73}
$$

$$
I_2 = (18, 21, 12, 1, 4, 14, 2, 5, 19, 17, 4, 6, 3).
\tag{4.74}
$$

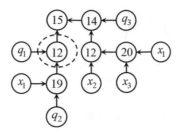

Fig. 4.5 IGP graph of the mathematical expression (4.75)

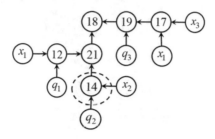

Fig. 4.6 IGP graph for the mathematical expression (4.76)

The codes correspond to the following mathematical expressions:

$$y_1 = A + B + AB + A^2 + B^2, \tag{4.75}$$

where

$$A = q_1 + x_1 + x_1 q_2 + x_1^2 + q_1^2,$$
$$B = (x_2 + x_1 + x_3 + x_1^2 + x_3^2 + x_2^2)^2 + q_3^2.$$

$$y_2 = CD + C^2 + D^2, \tag{4.76}$$

where

$$C = (x_1 + q_1 + x_1^2)(q_2^2 + x_2^2),$$
$$D = x_1 + x_1 x_3 + x_1^2 + x_3^2(1 + q_3) + (x_1 + x_1 x_3 + x_1^2 + x_3^2)^2.$$

Computational graphs of these expressions are shown in Figs. 4.5 and 4.6.
The nodes marked on the graphs are selected as crossover points.
Crossover subtrees have the following codes:

$$I_1(2) = (12, 1, 19, 4, 2),$$
$$I_2(5) = (14, 2, 5). \tag{4.77}$$

Here in brackets is the position number in the code with which the subsequence
begins.

The following codes of new mathematical expressions are obtained after
crossover and exchanging subcodes (4.77):

$$I_3 = (15, \underbrace{14, 2, 5}, 14, 12, 5, 20, 6, 4, 3), \tag{4.78}$$

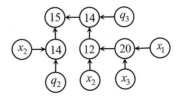

Fig. 4.7 IGP graph of the mathematical expression (4.80)

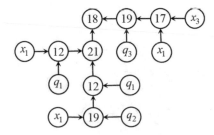

Fig. 4.8 IGP graph of the mathematical expression (4.81)

$$I_4 = (18, 21, 12, 1, 4, \underbrace{12, 1, 19, 4, 2}, 19, 17, 4, 6, 3). \tag{4.79}$$

The newly obtained codes correspond to the following mathematical expressions:

$$y_3 = E + F + EF + E^2 + F^2, \tag{4.80}$$

where

$$E = q_2^2 + x_2^2,$$
$$F = (x_2 + x_1 + x_3 + x_1^2 + x_3^2 + x_2^2)^2 + q_3^2.$$

$$y_4 = GH + G^2 + H^2 \tag{4.81}$$

where

$$G = (q_1 + x_1 + q_1^2)(q_1 + x_1 + x_1 q_2 + x_1^2 + q_1^2),$$
$$H = x_1 + x_1 x_3 + x_1^2 + x_3^2(1 + q_3) + (x_1 + x_1 x_3 + x_1^2 + x_3^2)^2.$$

The computational graphs of the obtained mathematical expressions are shown in Figs. 4.7 and 4.8.

4.5 Analytic Programming

Analytical programming [9, 10] is one more kind of symbolic regression methods. Elements of a mathematical expression in analytic programming are encoded with integers. All elements of the mathematical expression, together with arguments, are collected in one ordered set. Each number in the ordered set is the element code.

As a basic set of functions, a combined set of functions is used, in which a set of arguments of a mathematical expression is introduced

$$
\begin{aligned}
F = \{ & f_1 = q_1, \ldots, f_p = q_p, f_{p+1} = x_1, \ldots, f_{p+n} = x_n, \\
& f_{m_0+1} = f_{1,1}(z), \ldots, f_{m_0+m_1} = f_{1,m_1}(z), \\
& f_{m_0+m_1+1} = f_{2,1}(z_1, z_2), \ldots, f_{m_0+m_1+m_2} = f_{2,m_2}(z_1, z_2), \\
& \ldots \\
& f_{K+1} = f_{r,1}(z_1, \ldots, z_r), \ldots, f_{K+m_r} = f_{r,m_r}(z_1, \ldots, z_r) \},
\end{aligned}
\tag{4.82}
$$

where $m_0 = p + n$,

$$
K = \sum_{i=0}^{r-1} m_i.
$$

The mathematical expression is encoded by a sequence of element numbers from the combined set (4.82)

$$
C = (c_1, \ldots, c_L),
\tag{4.83}
$$

where c_j is the number of an element from the combined set (4.82), $j = 1, \ldots, L$.

In analytical programming, the prefix order of the elements is used. The function code in the record precedes the argument code. The length of the code record is limited.

Consider an Example

Suppose the following combined set of functions is defined:

$$
F = \{ q_1, q_2, x_1, x_2, -z, \sin(z), \cos(z), \exp(z), z_1 + z_1, z_1 z_2 \}.
\tag{4.84}
$$

and the following code of a mathematical expression is given:

$$
C_1 = (10, 9, 10, 1, 7, 3, 10, 2, 6, 4, 8, 5, 3).
\tag{4.85}
$$

Let us decode this code.
The first element $c_1^1 = 10$ corresponds to the multiplication function

$$
y_1 = z_1 z_2.
$$

Next element $c_1^2 = 9$ corresponds to the addition function

$$
y_1 = (z_1 + z_3) z_2.
$$

Then $c_1^3 = 10$, again multiplication

$$
y_1 = (z_1 z_4 + z_3) z_2.
$$

Next $c_1^4 = 1$ is the parameter q_1

$$
y_1 = (q_1 z_4 + z_3) z_2.
$$

$c_1^5 = 7$ is cos

$$y_1 = (q_1 \cos(z_1) + z_3)z_2.$$

$c_1^6 = 3$ is the variable x_1

$$y_1 = (q_1 \cos(x_1) + z_3)z_2.$$

Then again multiplication $c_1^7 = 10$

$$y_1 = (q_1 \cos(x_1) + z_1 z_3)z_2.$$

$c_1^8 = 2$ is the parameter q_2

$$y_1 = (q_1 \cos(x_1) + q_2 z_3)z_2.$$

$c_1^9 = 6$ is sin

$$y_1 = (q_1 \cos(x_1) + q_2 \sin(z_1))z_2.$$

$c_1^1 0 = 4$ is the variable x_2

$$y = (q_1 \cos(x_1) + q_2 \sin(x_2))z_2.$$

$c_1^{11} = 8$ is exp

$$y = (q_1 \cos(x_1) + q_2 \sin(x_2)) \exp(z_1).$$

$c_1^1 2 = 5$ corresponds to the sign change function

$$y = (q_1 \cos(x_1) + q_2 \sin(x_2)) \exp(-z_1).$$

$c_1^1 3 = 3$ is the variable x_1

$$y = (q_1 \cos(x_1) + q_2 \sin(x_2)) \exp(-x_1).$$

The correctness of the code is determined by the index of the code element. The index of the element i is determined from the relation

$$T(i) = 1 - i + \sum_{j=1}^{i} a_j, \tag{4.86}$$

where

$$a_j = \begin{cases} 0, \text{ if } a_j \le m_0, \\ k, \text{ if } \sum_{s=0}^{k} m_s < a_j \le \sum_{s=0}^{k+1} m_s, \ k = 0, \ldots, r-1. \end{cases} \tag{4.87}$$

The condition of the code correctness is defined by (3.10).

Crossover in analytic programming is accomplished by exchanging subcodes. Crossover points are randomly selected in each of the codes selected for crossover. From these points, sequences of subcodes are determined as independent codes, which start from the crossover point and satisfy the conditions for correct writing of the code. The last subcode element is determined by the zero value of the subcode element index. For example, in the code, the point α is selected with the element a_α. The indices of all elements of the subcode are calculated starting from the element α by the formula (4.86) with the replacement $i \leftarrow i - \alpha + 1$.

Consider an Example

Let the code of the second mathematical expression be given

$$C_2 = (10, 8, 10, 5, 1, 3, 7, 6, 10, 2, 4). \tag{4.88}$$

The code corresponds to the mathematical expression

$$y_2 = \exp(-q_1 x_1) \cos(\sin(q_2 x_2)). \tag{4.89}$$

Let in the code C_1 the crossover point be $\alpha = 3$ and in the code C_2, the crossing point be $\beta = 7$. Thus, the subcodes for crossover are

$$C_1(3) = (10, 1, 7, 3),$$
$$C_2(7) = (7, 6, 10, 2, 4).$$

Exchange the selected subcodes in the codes, and codes for new mathematical expressions are obtained

$$C_3 = (10, 9, \underbrace{7, 6, 10, 2, 4}, 10, 2, 6, 4, 8, 5, 3),$$
$$C_4 = (10, 8, 10, 5, 1, 3, \underbrace{10, 1, 7, 3}).$$

The resulting codes correspond to new mathematical expressions

$$y_3 = (\cos(\sin(q_2 x_2)) + q_2 \sin(x_2)) \exp(-x_1),$$
$$y_4 = \exp(-q_1 x_1) q_1 \cos(x_2).$$

Analytical programming has the shortest code to write a mathematical expression. The drawback of analytical programming is, as in genetic programming and inductive genetic programming, different lengths of codes of various mathematical expressions, and a change in the length of the code after the crossover operation.

4.6 Parse-Matrix Evolution

The parse-matrix evolution method [11] encodes function call commands as an integer vector of a certain length. The dimension of the function call vector is determined by the maximum number of arguments of the functions used. In this case, the method of parse-matrix evolution is similar to the method of Cartesian genetic programming. The elements of the function call vector include the code of the function, the codes of the function arguments, and the element for storing the calculation results. The dimension of the function call vector determines the number of columns in the parse-matrix. Unlike Cartesian genetic programming, the function call vector specifies the element code to store the results, and all function call vectors are combined into one parse-matrix.

To encode a mathematical expression by the parse-matrix, we define the basic ordered set of elementary functions

$$
\begin{aligned}
F = \{ & f_1 = f_{1,1}(z), \ldots, f_{m_1} = f_{1,m_1}(z), \\
& f_{m_1+1} = f_{2,1}(z_1,z_2), \ldots, f_{m_1+m_2} = f_{2,m_2}(z_1,z_2), \\
& \ldots \\
& f_{S_{r-1}+1} = f_{r,1}(z_1,\ldots z_r), \ldots, f_{S_r} = f_{r,m_r}(z_1,\ldots,z_r) \},
\end{aligned}
\tag{4.90}
$$

where

$$
S_k = \sum_{i=1}^{k} m_k, \ k = 1, \ldots, r.
$$

Define an ordered set of elements for storing the results of calculations

$$
S = (s_1, \ldots, s_v). \tag{4.91}
$$

Define an ordered united set of parameters, variables, and elements for storing the results of calculations. These elements are arguments of the mathematical expression and of functions from (4.90)

$$
\begin{aligned}
A = (& a_1 = q_1, \ldots, a_p = q_p, a_{p+1} = x_1, \ldots, a_{p+n} = x_n, \\
& a_{p+n+1} = s_1, \ldots, a_{p+n+v} = s_v).
\end{aligned}
\tag{4.92}
$$

The code of each element in the sets of elementary functions (4.90), arguments (4.92), and elements for storing the results of calculations (4.91) is determined by an ordinal number of the element by the formula

$$
c = n(C) - \left\lfloor \frac{|C|}{2} \right\rfloor, \tag{4.93}
$$

where $n(C)$ is an ordinal number of the element in the set C, $|C|$—is the cardinality of the set C, $C \in \{F, A, S\}$.

The code of the mathematical expression is written in the form of a parse-matrix with the dimension $L \times (r+2)$

$$
\mathbf{P} = [p_{i,j}], \ i = 1, \ldots, L, \ j = 1, \ldots, r+2. \tag{4.94}
$$

Each row of the parse-matrix (4.94) contains the code for calling a function from a set of elementary functions (4.90)

$$
p_{i,1} = n(F) - \left\lfloor \frac{|F|}{2} \right\rfloor, \tag{4.95}
$$

$$p_{i,r+2} = n(S) - \left\lfloor \frac{|S|}{2} \right\rfloor, \tag{4.96}$$

$$p_{i,j} = n(A) - \left\lfloor \frac{|A|}{2} \right\rfloor, \ j = 2, \dots, r+1. \tag{4.97}$$

The solution in PME symbolic regression method is searched using a genetic algorithm. When searching, all matrices of the set of possible solutions have the same dimension. For this purpose, some rows of parse-matrices have function calls that are not used for evaluation of the mathematical expression.

To perform the crossover operation, two parse-matrices are randomly selected

$$\mathbf{P}^\alpha = [p_{i,j}^\alpha], \ \mathbf{P}^\beta = [p_{i,j}^\beta], \ i = 1, \dots, L, \ j = 1, \dots, r+2. \tag{4.98}$$

Randomly select the crossover point $t \in \{1, \dots, L\}$ and exchange the matrix rows from t string to the end. Getting new parsing matrices

$$\mathbf{P}^\gamma = \begin{bmatrix} p_{1,1}^\alpha & \cdots & p_{1,r+2}^\alpha \\ \vdots & \ddots & \vdots \\ p_{t-1,1}^\alpha & \cdots & p_{t-1,r+2}^\alpha \\ p_{t,1}^\beta & \cdots & p_{t,r+2}^\beta \\ \vdots & \ddots & \vdots \\ p_{L,1}^\beta & \cdots & p_{L,r+2}^\beta \end{bmatrix}, \ \mathbf{P}^\delta = \begin{bmatrix} p_{1,1}^\beta & \cdots & p_{1,r+2}^\beta \\ \vdots & \ddots & \vdots \\ p_{t-1,1}^\beta & \cdots & p_{t-1,r+2}^\beta \\ p_{t,1}^\alpha & \cdots & p_{t,r+2}^\alpha \\ \vdots & \ddots & \vdots \\ p_{L,1}^\alpha & \cdots & p_{L,r+2}^\alpha \end{bmatrix}. \tag{4.99}$$

To perform the mutation operation, randomly select the mutation point $p_{a,b}$, $a \in \{1, \dots, L\}$, $b \in \{1, \dots, r+2\}$ and randomly generate a new value depending on the column b

$$p_{a,b} = \begin{cases} \mu - \left\lfloor \frac{|F|}{2} \right\rfloor, \ \mu \in \{1 \dots |F|\}, \text{ if } b = 1, \\ \mu - \left\lfloor \frac{|A|}{2} \right\rfloor, \ \mu \in \{1 \dots |A|\}, \text{ if } 1 < b \leq r+1, \\ \mu - \left\lfloor \frac{|S|}{2} \right\rfloor, \ \mu \in \{1 \dots |S|\}, \text{ if } b = r+2. \end{cases} \tag{4.100}$$

Consider an Example

Suppose the following mathematical expressions are given:

$$y_1 = \begin{cases} q_1 x_1 + x_2 \exp(-q_2 x_2) \cos(q_1 x_1), \text{ if } B < 0 \\ q_1 x_1 + x_2 \sin(q_2 x_2 + q_3), \text{ otherwise} \end{cases}, \tag{4.101}$$

$$y_2 = \begin{cases} \exp(-q_1 x_1), \text{ if } A \leq 0 \\ -\exp(-q_1 x_1), \text{ otherwise} \end{cases}, \tag{4.102}$$

where

$$B = x_1^2 - x_2^2,$$

$$A = \sin(\cos(q_1 x_1)) - \cos(\sin(q_2 x_2 + q_3)).$$

To encode the mathematical expressions, the following set of functions is used:

$$F = \{f_1 = -z, f_2 = \exp(z), f_3 = \sin(z), f_4 = \cos(z),$$
$$f_5 = z_1 + z_2, f_6 = z_1 z_2, f_7 = f_{3,1}(z_1, z_2, z_3),$$
$$f_8 = f_{3,2}(z_1, z_2, z_3)\}, \tag{4.103}$$

where

$$f_{3,1}(z_1, z_2, z_3) = \begin{cases} z_2, \text{ if } z_1 \leq 0 \\ z_3, \text{ otherwise} \end{cases},$$

$$f_{3,2}(z_1, z_2, z_3) = \begin{cases} z_3, \text{ if } z_1 \leq z_2 \\ -z_3, \text{ otherwise} \end{cases}.$$

Define the set for storing intermediate computations as

$$S = \{s_1, \ldots, s_{15}\}. \tag{4.104}$$

Define a set of arguments

$$A = \{a_1 = q_1, a_2 = q_2, a_3 = q_3, a_4 = x_1, a_5 = x_2,$$
$$a_6 = s_1, \ldots, a_{20} = s_{15}\}. \tag{4.105}$$

The maximum number of function arguments is $r = 3$, so the number of columns in the parse-matrix is $r + 2 = 5$.

Let us write sequentially the mathematical expressions (4.101), (4.102) using functions from the set F and encode them using PME.

$$q_1 x_1 : f_6(a_1, a_4), \ p_{1,1} = 6 - \left\lfloor \frac{|F|}{2} \right\rfloor = 6 - 4 = 2,$$

$$p_{1,2} = 1 - \left\lfloor \frac{|A|}{2} \right\rfloor = 1 - 10 = -9,$$

$$p_{1,3} = 4 - \left\lfloor \frac{|A|}{2} \right\rfloor = 4 - 10 = -6,$$

$p_{1,4}$ is not used and can be of any value,

$$p_{1,5} = 1 - \left\lfloor \frac{|S|}{2} \right\rfloor = 1 - 7 = -6.$$

As a result, the first row of the parse-matrix is obtained

$$\mathbf{p}^1 = [2\ -9\ -6\ *\ -6].$$

Perform the rest of the transformations.

$$\cos(q_1 x_1): s_2 = f_4(a_6),$$
$$\mathbf{p}^2 = [0\ -4\ *\ *\ -5],$$
$$q_2 x_2: s_3 = f_6(a_2, a_5),$$
$$\mathbf{p}^3 = [2\ -8\ -5\ *\ -4],$$
$$-q_2 x_2: s_4 = f_1(a_8),$$
$$\mathbf{p}^4 = [-3\ -2\ *\ *\ -3],$$
$$\exp(-q_2 x_2): s_5 = f_2(a_9),$$
$$\mathbf{p}^5 = [-2\ -1\ *\ *\ -2],$$
$$\exp(-q_2 x_2)\cos(q_1 x_1): s_6 = f_6(a_{10}, a_7),$$
$$\mathbf{p}^6 = [2\ 0\ -3\ *\ -1],$$
$$q_2 x_2 + q_3: s_7 = f_5(a_8, a_3),$$
$$\mathbf{p}^7 = [1\ -2\ -7\ *\ 0],$$
$$\sin(q_2 x_2 + q_3): s_8 = f_3(a_{12}),$$
$$\mathbf{p}^8 = [1\ 2\ *\ *\ 1],$$
$$x_1^2: s_9 = f_6(a_4, a_4),$$
$$\mathbf{p}^9 = [2\ -6\ -6\ *\ 2],$$
$$x_2^2: s_{10} = f_6(a_5, a_5),$$
$$\mathbf{p}^{10} = [2\ -5\ -5\ *\ 3],$$
$$-x_2^2: s_{11} = f_1(a_{15}),$$
$$\mathbf{p}^{11} = [-3\ 5\ *\ *\ 4],$$
$$x_1^2 - x_2^2: s_{12} = f_5(a_{14}, a_{16}),$$
$$\mathbf{p}^{12} = [1\ 4\ 6\ *\ 5],$$
$$f_{3,1}(x_1^2 - x_2^2, \exp(-q_1 x_1)\cos(q_1 x_1), \sin(q_2 x_2 + q_3)):$$
$$s_{13} = f_7(a_{17}, a_{11}, a_{13}),$$
$$\mathbf{p}^{13} = [3\ 7\ 1\ 3\ 6],$$
$$x_2 f_{3,1}(s_{12}, s_6, s_8): s_{14} = f_6(a_5, a_{18}),$$
$$\mathbf{p}^{14} = [2\ -5\ 8\ *\ 7],$$
$$y_1 = q_1 x_1 + f_{3,1}(s_{12}, s_6, s_8), s_{15} = f_5(a_7, a_{19}),$$
$$\mathbf{p}^{15} = [1\ -3\ 9\ *\ 8].$$

As a result, a parse-matrix of size 15×5 is obtained

$$
\mathbf{P}_1 =
\begin{bmatrix}
2 & -9 & -6 & * & -6 \\
0 & -4 & * & * & -5 \\
2 & -8 & -5 & * & -4 \\
-3 & -2 & * & * & -3 \\
-2 & -1 & * & * & -2 \\
2 & 0 & -3 & * & -1 \\
1 & -2 & -7 & * & 0 \\
-1 & 2 & * & * & 1 \\
2 & -6 & -6 & * & 2 \\
2 & -5 & -5 & * & 3 \\
-3 & 5 & * & * & 4 \\
1 & 4 & 6 & * & 5 \\
3 & 7 & 1 & 3 & 6 \\
2 & -5 & 8 & * & 7 \\
1 & -3 & 9 & * & 8
\end{bmatrix}.
\tag{4.106}
$$

The second mathematical expression (4.102) is coded by the following parse-matrix:

$$
\mathbf{P}_2 =
\begin{bmatrix}
2 & -9 & -6 & * & -6 \\
1 & -9 & -6 & * & -5 \\
-3 & -4 & * & * & -4 \\
-2 & -2 & * & * & -3 \\
0 & -4 & * & * & -2 \\
2 & -1 & 0 & * & -1 \\
-1 & 0 & * & * & 0 \\
-1 & -4 & * & * & -1 \\
2 & -1 & 3 & * & 2 \\
2 & -8 & -5 & * & 3 \\
1 & 5 & -7 & * & 4 \\
-1 & 6 & * & * & 5 \\
2 & -1 & 7 & * & 6 \\
0 & 7 & * & * & 7 \\
4 & 2 & 9 & -1 & 8
\end{bmatrix}.
\tag{4.107}
$$

In this matrix, the rows 2, 6, 9, 10, 13 describe function calls that are not used in the calculation of the mathematical expression (4.102). These lines are entered to preserve the dimension of the parse-matrix.

Define the crossover point, for example $t = 8$. After exchanging rows, new parse-matrices are received

$$
\mathbf{P}_3 = \begin{bmatrix}
2 & -9 & -6 & * & -6 \\
0 & -4 & * & * & -5 \\
2 & -8 & -5 & * & -4 \\
-3 & -2 & * & * & -3 \\
-2 & -1 & * & * & -2 \\
2 & 0 & -3 & * & -1 \\
1 & -2 & -7 & * & 0 \\
-1 & -4 & * & * & -1 \\
2 & -1 & 3 & * & 2 \\
2 & -8 & -5 & * & 3 \\
1 & 5 & -7 & * & 4 \\
-1 & 6 & * & * & 5 \\
2 & -1 & 7 & * & 6 \\
0 & 7 & * & * & 7 \\
4 & 2 & 9 & -1 & 8
\end{bmatrix}, \quad
\mathbf{P}_4 = \begin{bmatrix}
2 & -9 & -6 & * & -6 \\
1 & -9 & -6 & * & -5 \\
-3 & -4 & * & * & -4 \\
-2 & -2 & * & * & -3 \\
0 & -4 & * & * & -2 \\
2 & -1 & 0 & * & -1 \\
-1 & 0 & * & * & 0 \\
-1 & 2 & * & * & 1 \\
2 & -6 & -6 & * & 2 \\
2 & -5 & -5 & * & 3 \\
-3 & 5 & * & * & 4 \\
1 & 4 & 6 & * & 5 \\
3 & 7 & 1 & 3 & 6 \\
2 & -5 & 8 & * & 7 \\
1 & -3 & 9 & * & 8
\end{bmatrix}.
\tag{4.108}
$$

The matrices correspond to the following mathematical expressions:

$$
y_3 = \begin{cases} -q_2 x_2, & \text{if } q_2 x_2 + q_3 \leq \cos(\sin(q_2 x_2 + q_3)) \\ q_2 x_2, & \text{otherwise} \end{cases},
\tag{4.109}
$$

$$
y_4 = \begin{cases} q_1 + x_1 + x_2 \exp(-q_1 x_1) \cos(q_1 x_1), & \text{if } x_1^2 - x_2^2 \leq 0 \\ q_1 + x_1 + x_2, & \text{otherwise} \end{cases}.
\tag{4.110}
$$

4.7 Binary Complete Genetic Programming

Binary Complete Genetic Programming (BCGP) [12] is also a kind of genetic programming. BCGP represents a mathematical expression in the form of a binary tree and uses only functions with one and two arguments. In a computational tree, functions with two arguments are associated with tree nodes. Functions with one argument are associated with tree arcs.

A complete binary tree has a certain number of elements, depending on the number of levels. Each level, except for the last one, contains the same number of functions with one and two arguments. The last level contains the same number of functions with one argument and arguments of the mathematical expression, which are schematically associated with the leaves of the tree.

In BCGP, a mathematical expression is a sequential notation of function and argument codes. All codes are in a specific order by level. At each level, the codes of functions with one argument are firstly written, then functions with two arguments, and at the last level the codes of the arguments of the mathematical expression are written. The number of function arguments is determined by the ordinal number of the element in the code.

Define basic sets of elementary functions

- a set of functions with two arguments

$$F_2 = \{f_1 = f_{2,1}(z_1,z_2),\ldots,f_v = f_{2,v}(z_1,z_2)\}, \tag{4.111}$$

- a set of functions with one argument

$$G_1 = \{g_1 = f_{1,1}(z),\ldots,g_w = f_{1,w}(z)\}, \tag{4.112}$$

- and a set of arguments of a mathematical expression

$$A = \{a_1 = q_1,\ldots,a_p = q_p, a_{p+1} = x_1,\ldots,a_{n+p} = x_n, \\ a_{n+p+1} = e_1,\ldots,a_{n+p+v} = e_v\}, \tag{4.113}$$

where e_i is a unit element of the function with two arguments $f_{2,i}(z_1,z_2)$, $i = 1,\ldots,v$.

As seen, unit elements for functions with two arguments are added to the set of arguments. The unit element of a function with two arguments is such a value of one of the arguments that the result of evaluating the function is equal to the value of the other argument. The unit element for the addition function is 0, and for the multiplication function is 1.

Functions with two arguments are commutative

$$f_{2,i}(z_1,z_2) = f_{2,i}(z_2,z_1), \ i = 1,\ldots,v. \tag{4.114}$$

The initial level of the binary tree is zero level. Let there be L levels in the binary tree. At the $k < L$ level of the binary tree, there are 2^k functions with one argument and the same number of functions with two arguments. At the level $k = L$ there are 2^L functions with one argument and the same number of arguments of the mathematical expression, taking into account the unit elements for functions with two arguments. There are totally

$$N = 2\sum_{k=0}^{L} 2^k = 2(2^{L+1} - 1) \tag{4.115}$$

elements in the BCGP code with L levels.

Consider an Example

Let a mathematical expression be given

$$y_1 = (x_1^2 - x_2^2)\cos(q_1 x_1 + q_2) + x_1 x_2 \exp(-q_3 x_1). \tag{4.116}$$

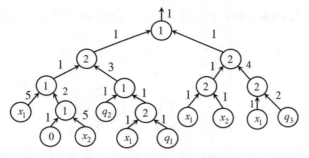

Fig. 4.9 A binary tree for the mathematical expression (4.116)

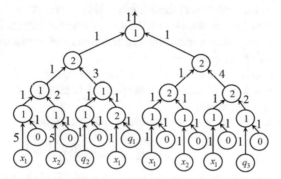

Fig. 4.10 A complete binary tree for the mathematical expression (4.116)

For encoding this mathematical expression, the following sets of functions and arguments are sufficient

$$F_2 = \{f_1 = z_1 + z_2, f_2 = z_1 z_2\}, \tag{4.117}$$

$$G_1 = \{g_1 = z, g_2 = -z, g_3 = \cos(z), g_4 = \exp(z), g_5 = z^2\}, \tag{4.118}$$

$$A = \{a_1 = q_1, a_2 = q_2, a_3 = q_3, a_4 = x_1, a_5 = x_2, a_6 = 0, a_7 = 1\}. \tag{4.119}$$

A binary computational tree for a mathematical expression is shown in Fig. 4.9.

The resulting binary tree (Fig. 4.9) is not complete and has no more than four levels. Add additional branches and nodes to the tree with functions with two arguments and single elements of these functions to get a complete binary tree.

The complete binary tree for the mathematical expression (4.116) is shown in Fig. 4.10.

BCGP code is an ordered set of integer numbers that indicate the numbers of elements from the sets of functions with one and two arguments and from the set of arguments of the encoded mathematical expression.

The BCGP code of the mathematical expression (4.116) is as follows:

$$
\begin{aligned}
C_1 = (&1, 1, \\
&1, 1, 2, 2, \\
&1, 3, 1, 4, \ 1, 1, 2, 2, \\
&1, 2, 1, 1, 1, 1, 1, 2, \ 1, 1, 1, 2, 1, 1, 1, 1, \\
&5, 1, 5, 1, 1, 1, 1, 1, 1, 1, 1, 1, 1, 1, 1, 1, \\
&4, 6, 5, 6, 2, 6, 4, 1, 4, 6, 5, 6, 4, 6, 3, 6).
\end{aligned}
\tag{4.120}
$$

To search for an optimal mathematical expression, the number of levels of a binary tree needs to be determined and codes of mathematical expressions are generated according to the structure of the complete binary tree. As a result, codes of all mathematical expressions in search process have the same length. In all codes, the arguments of a mathematical expression and functions with one and two arguments are at specific positions in the code.

To perform the crossover operation, one crossover point is chosen for both selected possible solutions. Crossover is performed conventionally by exchanging codes after the crossover point.

To demonstrate the crossover operation, consider the second mathematical expression

$$
y_2 = \cos(\exp(-(q_1 x_1^2 + q_2 x_2^2))).
\tag{4.121}
$$

The BCGP code of the mathematical expression (4.121) is as follows:

$$
\begin{aligned}
C_2 = (&3, 1, \\
&4, 1, \ 1, 1, \\
&2, 1, 1, 1 \ 1, 1, 1, 1, \\
&1, 1, 1, 1, 1, 1, 1, 1, \ 2, 2, 1, 1, 1, 1, 1, 1, \\
&5, 1, 5, 1, 1, 1, 1, 1, 1, 1, 1, 1, 1, 1, 1, 1, \\
&4, 1, 5, 2, 6, 6, 6, 6, 6, 6, 6, 6, 6, 6, 6, 6).
\end{aligned}
\tag{4.122}
$$

Randomly choose the crossover point $s \in \{1, \dots, 62\}$. Suppose $s = 26$.

Exchange the elements of the codes from the crossover point and get new codes of mathematical expressions

$$
\begin{aligned}
C_3 = (&1, 1, \\
&1, 1, 2, 2, \\
&1, 3, 1, 4, \ 1, 1, 2, 2, \\
&1, 2, 1, 1, 1, 1, 1, 2, \ 1, 1, 1, 2, 1, 1, 1, 1, \\
&5, 1, 5, 1, 1, 1, 1, 1, 1, 1, 1, 1, 1, 1, 1, 1, \\
&4, 1, 5, 2, 6, 6, 6, 6, 6, 6, 6, 6, 6, 6, 6, 6).
\end{aligned}
\tag{4.123}
$$

$$
\begin{aligned}
C_4 = (&3, 1, \\
&4, 1, \ 1, 1, \\
&2, 1, 1, 1, \ 1, 1, 1, 1, \\
&1, 1, 1, 1, 1, 1, 1, 1, \ 2, 2, 1, 1, 1, 1, 1, 1, \\
&5, 1, 5, 1, 1, 1, 1, 1, 1, 1, 1, 1, 1, 1, 1, 1, \\
&4, 6, 5, 6, 2, 6, 4, 1, 4, 6, 5, 6, 4, 6, 3, 6).
\end{aligned}
\tag{4.124}
$$

The resulting codes correspond to the following mathematical expressions:

$$y_3 = q_1 x_1^2 - q_2 x_2^2, \tag{4.125}$$

$$y_4 = \exp(x_1 + q_1 + q_2) + 2x_1 + q_3. \tag{4.126}$$

4.8 Network Operator Method

The network operator method [13–15] encodes a mathematical expression in the form of a directed graph. The method only uses functions with one and two arguments. In the directed graph of the network operator, arguments of the mathematical expression are associated with source nodes, functions with one argument are associated with graph arcs, functions with two arguments are associated with other nodes in the graph. Any nodes of the graph can be defined as outputs, and the results of calculations in them will be the values of the components of the output vector. Sink-nodes in a graph are outputs of the mathematical expression.

To encode a mathematical expression by the network operator method, three ordered sets are used

- and a set of arguments of the mathematical expression, or parameters and variables of the mathematical expression

$$F_0 = \{f_{0,1} = q_1, \ldots, f_{0,p} = q_p, f_{0,p+1} = x_1, \ldots, f_{0,n+p} = x_n\}, \tag{4.127}$$

- a set of functions with one argument

$$F_1 = \{f_{1,1} = z, f_{1,2}(z), \ldots, f_{1,w}(z)\}, \tag{4.128}$$

- a set of functions with two arguments

$$F_2 = \{f_{2,1}(z_1, z_2), \ldots, f_{2,v}(z_1, z_2)\}. \tag{4.129}$$

A set of functions with one argument F_1 must necessarily contain the identical function

$$f_{1,1} = z.$$

Functions with two arguments must be associative, commutative, and have a unit element.

To build a computational oriented graph of the network operator, it is necessary to represent a mathematical expression in the form of a composition of functions. The first function in the composition should be a function with two arguments. Further in the composition notation, functions with one argument alternate with functions with two arguments.

Also, when building a network operator graph, the following rules must also be observed:

- Arguments of functions with one argument are only functions with two arguments or elements of a set of parameters and variables,
- Arguments of a function with two arguments are only functions with one argument or its unit element,
- Functions with two arguments must not include functions with the same argument as arguments.

To satisfy these requirements for constructing a network operator graph, additional functions with two arguments and with a unit element of this function as the second argument are introduced. A unit element is not indicated on the graph. If a graph node contains one arc, then the second argument associated with this node is a unit element, therefore a function with two arguments associated with a node with only one arc does not change the input value.

Consider an Example

Let a mathematical expression be given

$$y = (x_1^2 - x_2^2)\cos(q_1 x_1 + q_2) + x_1 x_2 \exp(-q_3 x_1). \qquad (4.130)$$

To encode this mathematical, the following sets of arguments and functions are needed

$$F_0 = \{x_1, x_2, q_1, q_2, q_3\}, \qquad (4.131)$$

$$F_1 = \{f_{1,1}(z) = z, f_{1,2}(z) = -z, f_{1,3}(z) = \cos(z), \\ f_{1,4}(z) = \exp(z), f_{1,5}(z) = z^2\}, \qquad (4.132)$$

$$F_2 = \{f_{2,1}(z_1, z_2) = z_1 + z_2, f_{2,2}(z_1, z_2) = z_1 z_2\}. \qquad (4.133)$$

The network operator graph of the mathematical expression (4.130) is shown in Fig. 4.11.

In the Fig. 4.11 the nodes contain the numbers of functions with two arguments, the numbers of functions with one argument are indicated next to the arcs, the arguments of the mathematical expression are indicated in the source nodes, and the numbers of the nodes are indicated in the upper parts of the nodes. Node are numbered according to the rules of topological sorting, the number of the node from which the arc exits is less than the number of the node where the arc enters.

In the computer memory, the graph of the network operator is represented as an integer square matrix of the network operator, which is constructed from the adjacency matrix of the graph. In the graph of the network operator, instead of 1 in the adjacency matrix, the numbers of functions with one argument are indicated, the diagonal contains the numbers of functions with two arguments.

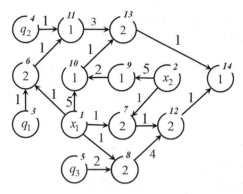

Fig. 4.11 A graph of the network operator for the mathematical expression (4.130)

The network operator matrix for the graph of the mathematical expression (4.130) has the following form:

$$\Psi = \begin{bmatrix} 0&0&0&0&0&1&1&1&0&5&0&0&0&0 \\ 0&0&0&0&0&0&1&0&5&0&0&0&0&0 \\ 0&0&0&0&0&1&0&0&0&0&0&0&0&0 \\ 0&0&0&0&0&0&0&0&0&0&1&0&0&0 \\ 0&0&0&0&0&0&0&2&0&0&0&0&0&0 \\ 0&0&0&0&0&2&0&0&0&0&1&0&0&0 \\ 0&0&0&0&0&0&2&0&0&0&0&1&0&0 \\ 0&0&0&0&0&0&0&2&0&0&0&4&0&0 \\ 0&0&0&0&0&0&0&0&1&2&0&0&0&0 \\ 0&0&0&0&0&0&0&0&0&1&0&0&1&0 \\ 0&0&0&0&0&0&0&0&0&1&0&3&0 \\ 0&0&0&0&0&0&0&0&0&0&2&0&1 \\ 0&0&0&0&0&0&0&0&0&0&0&2&1 \\ 0&0&0&0&0&0&0&0&0&0&0&0&1 \end{bmatrix}.$$ (4.134)

The complexity of constructing a graph of a network operator by a mathematical expression is compensated by the fact that such a construction is not required to be performed when searching for a mathematical expression. Any integer square upper-triangular matrix in which zero columns correspond to source nodes, and in the remaining columns and in all rows there are non-diagonal non-zero elements whose values correspond to the numbers of functions with one argument, and on the diagonal in non-zero columns there are numbers of functions with two arguments, is the network operator matrix that describes some computational graph.

Consider an example of calculating a mathematical expression using the graph of a network operator.

To store the results of intermediate calculations, a vector of nodes is introduced, the dimension of which is equal to the number of rows of the network operator matrix.

Initialize the vector of nodes and put the values of the arguments and the unit elements of the functions with two arguments in its corresponding components

$$\mathbf{z}^{(0)} = [x_1 \; x_2 \; q_1 \; q_2 \; q_3 \; 1 \; 1 \; 1 \; 0 \; 0 \; 0 \; 1 \; 1 \; 0]^T.$$

Scan the matrix (4.134) row by row and find non-zero non-diagonal elements $\psi_{i,j} \neq 0$, $i = 1, \ldots, 13$, $j = i+1, \ldots, 14$.

$$\psi_{1,6} = 1, \; z_6^{(1)} = f_{2,2}(z^{(0)6}, f_{1,1}(x_1)) = 1 \cdot x_1 = x_1,$$
$$\psi_{1,7} = 1, \; z_7^{(1)} = f_{2,2}(z_7^{(0)}, f_{1,1}(x_1)) = 1 \cdot x_1 = x_1,$$
$$\psi_{1,8} = 1, \; z_8^{(1)} = f_{2,2}(z_8^{(0)}, f_{1,1}(x_1)) = 1 \cdot x_1 = x_1,$$
$$\psi_{1,10} = 5, \; z_{10}^{(1)} = f_{2,2}(z_{10}^{(0)}, f_{1,5}(x_1)) = 1 \cdot x_1^2 = x_1^2,$$
$$z_i^{(1)} = z_i^{(0)}, \; i = 1, \ldots, 5, 9, 11, \ldots, 14,$$

$$\psi_{2,7} = 1, z_7^{(2)} = f_{2,2}(z_7^{(1)}, f_{1,1}(x_2)) = x_1 \cdot x_2 = x_1 x_2,$$

$$\psi_{2,9} = 5, \; z_9^{(2)} = f_{2,2}(z_9^{(1)}, f_{1,5}(x_2)) = 1 \cdot x_2^2 = x_2^2,$$
$$z_i^{(2)} = z_i^{(1)}, \; i = 1, \ldots, 6, 8, 10, \ldots, 14,$$

$$\psi_{3,6} = 1, z_6^{(3)} = f_{2,2}(z_6^{(2)}, f_{1,1}(q_1)) = x_1 \cdot q_1 = x_1 q_1,$$
$$z_i^{(3)} = z_i^{(2)}, \; i = 1, \ldots, 5, 7, \ldots, 14,$$

$$\psi_{4,11} = 1, z_{11}^{(4)} = f_{2,1}(z_{11}^{(3)}, f_{1,1}(q_2)) = 0 + q_2 = q_2,$$
$$z_i^{(4)} = z_i^{(3)}, \; i = 1, \ldots, 10, 12, 13, 14,$$

$$\psi_{5,8} = 2, z_8^{(5)} = f_{2,2}(z_8^{(4)}, f_{1,2}(q_3)) = 1 \cdot (-q_3) = -q_3,$$
$$z_i^{(5)} = z_i^{(4)}, \; i = 1, \ldots, 7, 9, \ldots, 14,$$

$$\psi_{6,11} = 2, z_{11}^{(6)} = f_{2,1}(z_{11}^{(5)}, f_{1,1}(z_6^{(5)})) = q_2 + x_1 q_1,$$
$$z_i^{(6)} = z_i^{(5)}, \; i = 1, \ldots, 10, 12, 13, 14,$$

$$\psi_{7,12} = 1, z_{12}^{(7)} = f_{2,2}(z_{12}^{(6)}, f_{1,1}(z_7^{(6)})) = x_1 x_2,$$
$$z_i^{(7)} = z_i^{(6)}, \; i = 1, \ldots, 11, 13, 14,$$

$$\psi_{8,12} = 4, z_{12}^{(8)} = f_{2,2}(z_{12}^{(7)}, f_{1,4}(z_8^{(7)})) = \exp(-x_1 q_3),$$
$$z_i^{(8)} = z_i^{(7)}, \; i = 1, \ldots, 11, 13, 14,$$

$$\psi_{9,10} = 2, z_{10}^{(9)} = f_{2,1}(z_{10}^{(8)}, f_{1,2}(z_9^{(8)})) = x_1^2 - x_2^2,$$
$$z_i^{(9)} = z_i^{(8)}, \; i = 1, \ldots, 9, 11, 12, 13, 14,$$

$$\psi_{10,13} = 1, z_{13}^{(10)} = f_{2,2}(z_{13}^{(9)}, f_{1,1}(z_{10}^{(9)})) = x_1^2 - x_2^2$$
$$z_i^{(10)} = z_i^{(9)}, \; i = 1, \ldots, 12, 14,$$

$$\psi_{11,13} = 3, z_{13}^{(11)} = f_{2,2}(z_{13}^{(10)}, f_{1,2}(z_{11}^{(10)})) =$$
$$(x_1^2 - x_2^2)\cos(x_1 q_1 + q_2)$$
$$z_i^{(11)} = z_i^{(10)}, \ i = 1, \ldots, 12, 14,$$

$$\psi_{12,14} = 1, z_{14}^{(12)} = f_{2,1}(z_{14}^{(11)}, f_{1,1}(z_{12}^{(11)})) = x_1 x_2 \exp(-x_1 q_3)$$
$$z_i^{(12)} = z_i^{(11)}, \ i = 1, \ldots, 13,$$

$$\psi_{13,14} = 1, z_{14}^{(13)} = f_{2,1}(z_{14}^{(12)}, f_{1,1}(z_{13}^{(12)})) =$$
$$(x_1^2 - x_2^2)\cos(x_1 q_1 + q_2) + x_1 x_2 \exp(-x_1 q_3)$$
$$z_i^{(13)} = z_i^{(12)}, \ i = 1, \ldots, 13.$$

As soon as all the rows of the matrix of the network operator are scanned, the last element of the vector of nodes contains the result of calculating the mathematical expression encoded by the network operator.

The network operator method is the first symbolic regression method, where the principle of small variations of basic solutions was applied.

So, the genetic operations are performed not directly on the network operator matrix, but on the set of variations.

According to this principle only one basic solution is coded in the form of network operator matrix. Other possible solutions are presented as small variations of this basic solution. For coding of every possible solution, the set of small variation vectors is used.

Each small variation vector includes four integer components

$$\mathbf{w} = [w_1 \ w_2 \ w_3 \ w_4]^T, \tag{4.135}$$

where w_1 is a type of variation, w_2 is a row number of the network operator matrix, w_3 is a column number of the network operator matrix, w_4 is a new value of the element in the network operator matrix.

The following types of small variations are used for the network operator matrix:

- $w_1 = 0$ is a replacement of a non-zero non-diagonal element,
- $w_1 = 1$ is a replacement of a non-zero diagonal element,
- $w_1 = 2$ is a replacement of zero non-diagonal element by a non-zero element,
- $w_1 = 3$ is a replacement of non-zero non-diagonal element by a zero element.

Variations are performed while maintaining the correctness condition for the network operator matrix

$$\forall i \ \exists \psi_{i,j} \neq 0, \ i \neq j \ \text{ and}$$
$$\forall j \notin \{s_1, \ldots, s_{N+P}\}, \ \exists \psi_{k,j} \neq 0, \ k \neq j. \tag{4.136}$$

This condition (4.136) of the correctness of the network operator matrix means the absence of such rows and columns that do not correspond to the source nodes, in which all non-diagonal elements are zero.

Consider an Example

Let a small variation vector be

$$\mathbf{w} = [2\ 8\ 10\ 3]^T.\tag{4.137}$$

After this small variation, the network operator matrix (4.134) has the following form:

$$\mathbf{w} \circ \Psi = \begin{bmatrix} 0\,0\,0\,0\,0\,1\,1\,1\,0\,5\,0\,0\,0\,0 \\ 0\,0\,0\,0\,0\,0\,1\,0\,5\,0\,0\,0\,0\,0 \\ 0\,0\,0\,0\,0\,1\,0\,0\,0\,0\,0\,0\,0\,0 \\ 0\,0\,0\,0\,0\,0\,0\,0\,0\,1\,0\,0\,0 \\ 0\,0\,0\,0\,0\,0\,0\,2\,0\,0\,0\,0\,0\,0 \\ 0\,0\,0\,0\,0\,2\,0\,0\,0\,0\,1\,0\,0\,0 \\ 0\,0\,0\,0\,0\,0\,2\,0\,0\,0\,0\,1\,0\,0 \\ 0\,0\,0\,0\,0\,0\,0\,2\,0\,3\,0\,4\,0\,0 \\ 0\,0\,0\,0\,0\,0\,0\,0\,1\,2\,0\,0\,0\,0 \\ 0\,0\,0\,0\,0\,0\,0\,0\,1\,0\,0\,1\,0 \\ 0\,0\,0\,0\,0\,0\,0\,0\,0\,1\,0\,3\,0 \\ 0\,0\,0\,0\,0\,0\,0\,0\,0\,0\,2\,0\,1 \\ 0\,0\,0\,0\,0\,0\,0\,0\,0\,0\,0\,2\,1 \\ 0\,0\,0\,0\,0\,0\,0\,0\,0\,0\,0\,0\,1 \end{bmatrix}.\tag{4.138}$$

This new network operator matrix codes the following mathematical expression:

$$\tilde{y} = (x_1^2 - x_2^2 + \cos(-q_3 x_1))\cos(x_1 q_1 + q_2) + x_1 x_2 \exp(-x_1 q_3).\tag{4.139}$$

Now let us consider genetic operations in the NOP method.

Each possible solution is coded by the set of small variation vectors

$$W_k = (\mathbf{w}^{k,1}, \ldots, \mathbf{w}^{k,d}),\tag{4.140}$$

where d is a number of small variations or a depth of variations.

The crossover is performed for the sets of small variations vectors. For this two possible solutions are selected

$$\begin{aligned} W_\alpha &= (\mathbf{w}^{\alpha,1}, \ldots, \mathbf{w}^{\alpha,d}), \\ W_\beta &= (\mathbf{w}^{\beta,1}, \ldots, \mathbf{w}^{\beta,d}). \end{aligned}\tag{4.141}$$

Randomly determine a crossover point

$$c \in \{1, \ldots, d\}.\tag{4.142}$$

Two new possible solutions are obtained by exchanging variation vectors after the crossover point

$$\begin{aligned} W_\gamma &= (\mathbf{w}^{\alpha,1}, \ldots, \mathbf{w}^{\alpha,c}, \mathbf{w}^{\beta,c+1}, \ldots, \mathbf{w}^{\beta,d}), \\ W_\delta &= (\mathbf{w}^{\beta,1}, \ldots, \mathbf{w}^{\beta,c}, \mathbf{w}^{\alpha,c+1}, \ldots, \mathbf{w}^{\alpha,d}). \end{aligned}\tag{4.143}$$

Mutation is performed for new possible solutions (4.143) by randomly generating new variation vectors in a randomly selected position.

4.9 Variational Symbolic Regression Methods

The principle of small variations of the basic solution [16], implemented in the network operator method, can be applied in other symbolic regression methods. All methods that use the principle of small variations include the word "variational" in the name. It is not a whim of the authors to apply the principle of small variations to different symbolic regression methods. The experience of studying the application of symbolic regression methods for solving complex control problems has shown that only variational methods of symbolic regression allow to obtain acceptable solutions. A detailed study of this issue is probably still ahead, and it is connected with the explanation of the global reasons for the success of evolutionary computations.

Why is the technology of small variations in possible solutions, moreover, performed with the use of a random number generator, better than an ordinary random search? Our hypothesis about the reason for the successful application of small variations in possible solutions in optimization problems is that these small variations must have a certain property. This property depends both on the algorithm itself or the parameters of the algorithm, and on the problem for the solution of which it is applied. Let us call this property the inheritance property.

Definition 4.1. Small variations of possible solutions that are used in the algorithm for finding the optimal value of the objective function possess the inheritance property, if after applying small variations to the set of possible solutions, there is always a non-zero constant part of possible solutions that have objective function values that differ from the values of the objective function for the same possible solutions before small variations by no more than a certain amount, and this does not depend on the value of the objective function.

Obviously, when in the process of searching for the optimal solution, some good possible solutions have already been found, which values of the objective function are close to the optimal, then it is unlikely to find a better possible solution by the random search. The probability of finding a better solution by the random search decreases with the improvement of the found possible solution. The better a possible solution found, the less likely is to find a better solution by the random search. And in the case of the inheritance property, with small variations of the already found good possible solutions, we always get a part of possible solutions in which the values of the objective function are in the vicinity of the already found good values. This means that the probability of finding a solution among the inheritors that will give a better value of the objective function becomes higher than that of the random search.

It is possible that the complex crossover operations used in symbolic regression methods do not have the property of inheritance. This leads to the fact that the search for the optimal solution becomes random.

Let us illustrate this in an example. In the classical genetic algorithm, which searches for the optimal solution in the form of a real vector on the space of Gray codes, after performing selection, crossover, and mutation for several generations, the set of codes of possible solutions turns out to be similar to each other. Ideally, when a global minimum is found, all possible solutions are the same. Crossing the same codes gives the same codes. In genetic programming, when crossing the same trees, due to different crossover points, we get new trees that are completely different from the trees of the parents.

Now consider several examples of applying the principle of small variations for various well-known symbolic regression methods. The application of this principle to each specific method requires the definition of the small variation types and the ways of their coding.

4.9.1 Variational Genetic Programming

Define small variations for genetic programming code [17].

- 1—change of the second component of the function code vector, while the value of the second component indicates the index of the element from the set given by the first component;
- 2—removal of the function with one argument;
- 3—insertion of a function with one argument;
- 4—increasing the value of the first component of the function vector code, while the vector of the argument code is inserted after the code of the function;
- 5—decreasing the value of the first component of the function vector code by one, while deleting the first argument code encountered after the variable code.

If some contradiction arises when performing a variation, then the small variation is not performed.

Small variation for GP can be described by a vector of variation of tree components

$$\mathbf{w} = [w_1 \ w_2 \ w_3]^T, \tag{4.144}$$

where w_1 is a type of variation, $w_1 \in \{1, \ldots, 5\}$, w_2 is an index of variable element, w_3 is a value of new element.

Consider an Example

Let the mathematical expression be (4.6)

$$y_0 = \exp(-ax_1)\cos(bx_2 + c),$$

and its GP code (4.10) is

$$S_0 = \left(\begin{bmatrix} 2 \\ 2 \end{bmatrix}, \begin{bmatrix} 1 \\ 2 \end{bmatrix}, \begin{bmatrix} 2 \\ 2 \end{bmatrix}, \begin{bmatrix} 1 \\ 1 \end{bmatrix}, \begin{bmatrix} 0 \\ 3 \end{bmatrix}, \begin{bmatrix} 0 \\ 1 \end{bmatrix}, \begin{bmatrix} 1 \\ 3 \end{bmatrix}, \right.$$

$$\left. \begin{bmatrix} 2 \\ 1 \end{bmatrix}, \begin{bmatrix} 2 \\ 2 \end{bmatrix}, \begin{bmatrix} 0 \\ 2 \end{bmatrix}, \begin{bmatrix} 0 \\ 4 \end{bmatrix}, \begin{bmatrix} 0 \\ 5 \end{bmatrix} \right).$$

The code uses the following sets of elementary functions and arguments:

$$\begin{aligned} F_0 &= \{f_{0,1} = x_1, f_{0,2} = x_2, f_{0,3} = a, f_{0,4} = b, f_{0,5} = c\}, \\ F_1 &= \{f_{1,1}(z) = -z, f_{1,2}(z) = \exp(z), f_{1,3}(z) = \cos(z), \\ &\quad f_{1,4}(z) = \sin(z), f_{1,5}(z) = z^2\}, \\ F_2 &= \{f_{2,1}(z_1, z_2) = z_1 + z_2, f_{2,2}(z_1, z_2) = z_1 z_2\}. \end{aligned} \tag{4.145}$$

We added to these sets (4.145) some new functions with one argument that were not used in the mathematical expression (4.6) in order to demonstrate variations further.

Let S_0 be a basic solution.

Let possible solutions be set in the form of ordered sets of small variation vectors.

$$W_1 = (w^{1,1}, \dots, w^{1,4}) = \left(\begin{bmatrix} 1 \\ 3 \\ 1 \end{bmatrix}, \begin{bmatrix} 3 \\ 10 \\ 3 \end{bmatrix}, \begin{bmatrix} 4 \\ 2 \\ 1 \end{bmatrix}, \begin{bmatrix} 5 \\ 1 \\ 2 \end{bmatrix} \right), \tag{4.146}$$

$$W_2 = (w^{2,1}, \dots, w^{2,4}) = \left(\begin{bmatrix} 1 \\ 1 \\ 1 \end{bmatrix}, \begin{bmatrix} 3 \\ 4 \\ 4 \end{bmatrix}, \begin{bmatrix} 3 \\ 2 \\ 5 \end{bmatrix}, \begin{bmatrix} 2 \\ 9 \\ 3 \end{bmatrix} \right). \tag{4.147}$$

Codes of mathematical expressions of these possible solutions are

$$S_1 = w^{1,4} \circ w^{1,3} \circ w^{1,2} \circ w^{1,1} \circ S_0 = \left(\begin{bmatrix} 1 \\ 2 \end{bmatrix}, \begin{bmatrix} 2 \\ 2 \end{bmatrix}, \begin{bmatrix} 2 \\ 1 \end{bmatrix}, \begin{bmatrix} 1 \\ 1 \end{bmatrix}, \right.$$

$$\left. \begin{bmatrix} 0 \\ 3 \end{bmatrix}, \begin{bmatrix} 0 \\ 1 \end{bmatrix}, \begin{bmatrix} 1 \\ 3 \end{bmatrix}, \begin{bmatrix} 2 \\ 1 \end{bmatrix}, \begin{bmatrix} 2 \\ 2 \end{bmatrix}, \begin{bmatrix} 1 \\ 3 \end{bmatrix}, \begin{bmatrix} 0 \\ 2 \end{bmatrix}, \begin{bmatrix} 0 \\ 4 \end{bmatrix}, \begin{bmatrix} 0 \\ 5 \end{bmatrix} \right), \tag{4.148}$$

$$S_2 = w^{2,4} \circ w^{2,3} \circ w^{2,2} \circ w^{2,1} \circ S_0 = \left(\begin{bmatrix} 2 \\ 1 \end{bmatrix}, \begin{bmatrix} 1 \\ 5 \end{bmatrix}, \begin{bmatrix} 1 \\ 2 \end{bmatrix}, \begin{bmatrix} 2 \\ 2 \end{bmatrix}, \right.$$

$$\left. \begin{bmatrix} 1 \\ 4 \end{bmatrix}, \begin{bmatrix} 1 \\ 1 \end{bmatrix}, \begin{bmatrix} 0 \\ 3 \end{bmatrix}, \begin{bmatrix} 0 \\ 1 \end{bmatrix}, \begin{bmatrix} 2 \\ 1 \end{bmatrix}, \begin{bmatrix} 2 \\ 2 \end{bmatrix}, \begin{bmatrix} 0 \\ 2 \end{bmatrix}, \begin{bmatrix} 0 \\ 4 \end{bmatrix}, \begin{bmatrix} 0 \\ 5 \end{bmatrix} \right). \tag{4.149}$$

To obtain variation of codes, the first variation vector $w^{1,1} = [1 \ 3 \ 1]^T$ is taken. The first component is $w_1^{1,1} = 1$, this means that it is necessary to change the second component in the code vector $w_2^{1,1} = 3$ for value $w_2^{1,1} = 1$. Result is $s^{1,3} = [2 \ 1]^T$. Further all other variations are made.

These codes of possible solutions correspond to mathematical expressions

$$y_1 = \exp((-a+x_1)\cos(b\cos(x_2)+c)),$$

$$y_2 = \exp^2(x_1\sin(-a))+bx_2+c.$$

Now, let us perform crossover for the possible solutions given in the form of variation vector sets (4.146), (4.147).

Select a crossover point $c \in \{1,\ldots,4\}$. For example, $c = 3$.

New possible solutions are the following:

$$W_3 = (\mathbf{w}^{1,1}, \mathbf{w}^{1,2}, \mathbf{w}^{2,3}, \mathbf{w}^{2,4}),$$
$$W_4 = (\mathbf{w}^{2,1}, \mathbf{w}^{2,2}, \mathbf{w}^{1,3}, \mathbf{w}^{1,4}).$$

Applying small variations to the basic solution, the following codes are obtained:

$$S_3 = \mathbf{w}^{3,4} \circ \mathbf{w}^{3,3} \circ \mathbf{w}^{3,2} \circ \mathbf{w}^{3,1} \circ S_0 = \left(\begin{bmatrix} 2 \\ 2 \end{bmatrix}, \begin{bmatrix} 1 \\ 5 \end{bmatrix}, \begin{bmatrix} 1 \\ 2 \end{bmatrix}, \begin{bmatrix} 2 \\ 1 \end{bmatrix}, \right.$$

$$\left. \begin{bmatrix} 1 \\ 1 \end{bmatrix}, \begin{bmatrix} 0 \\ 3 \end{bmatrix}, \begin{bmatrix} 0 \\ 1 \end{bmatrix}, \begin{bmatrix} 1 \\ 3 \end{bmatrix}, \begin{bmatrix} 2 \\ 1 \end{bmatrix}, \begin{bmatrix} 2 \\ 2 \end{bmatrix}, \begin{bmatrix} 1 \\ 3 \end{bmatrix}, \begin{bmatrix} 0 \\ 2 \end{bmatrix}, \begin{bmatrix} 0 \\ 4 \end{bmatrix}, \begin{bmatrix} 0 \\ 5 \end{bmatrix} \right),$$

$$S_4 = \mathbf{w}^{4,4} \circ \mathbf{w}^{4,3} \circ \mathbf{w}^{4,2} \circ \mathbf{w}^{4,1} \circ S_0 = \left(\begin{bmatrix} 1 \\ 5 \end{bmatrix}, \begin{bmatrix} 2 \\ 2 \end{bmatrix}, \begin{bmatrix} 2 \\ 2 \end{bmatrix}, \begin{bmatrix} 1 \\ 4 \end{bmatrix}, \right.$$

$$\left. \begin{bmatrix} 1 \\ 1 \end{bmatrix}, \begin{bmatrix} 0 \\ 3 \end{bmatrix}, \begin{bmatrix} 0 \\ 1 \end{bmatrix}, \begin{bmatrix} 1 \\ 3 \end{bmatrix}, \begin{bmatrix} 2 \\ 1 \end{bmatrix}, \begin{bmatrix} 2 \\ 2 \end{bmatrix}, \begin{bmatrix} 0 \\ 2 \end{bmatrix}, \begin{bmatrix} 0 \\ 4 \end{bmatrix}, \begin{bmatrix} 0 \\ 5 \end{bmatrix} \right).$$

New codes correspond the following mathematical expressions:

$$y_3 = \exp(-a+x_1)+\cos(b\cos(x_2)+c),$$

$$y_4 = (x_1\sin(-a)\cos(bx_2+c))^2.$$

4.9.2 Variational Analytic Programming

The analytic programming is a kind of genetic programming. The difference in the codes of mathematical expressions is that in the analytical programming the number of function arguments is not specified directly, and is determined by the number of functions. It follows from this that small variations and the coding vector of these variations coincide with the variational genetic programming.

A vector of tree components is used to encode small variation [18]

$$\mathbf{w} = [w_1 \; w_2 \; w_3]^T, \tag{4.150}$$

where w_1 is a type of small variation, w_2 is a code position number, w_3 is a new code value.

The following variations are possible:

- $w_1 = 1$—change the function number while maintaining the number of arguments;
- $w_1 = 2$—remove the function number from the code, if this is a function with one argument;
- $w_1 = 3$—insert into the code a function with one argument;
- $w_1 = 4$—change the function number so that the number of function arguments increases by one, while after this function the code of the element from the set of arguments is added;
- $w_1 = 5$—change the number of the function so that the number of function arguments decreases by one, while the first argument code that was found after the code of the changed function is removed from the code.

Consider an Example

Let the analytical programming code of the mathematical expression be given (4.85)

$$C = (10, 9, 10, 1, 7, 3, 10, 2, 6, 4, 8, 5, 3).$$

A combined set of functions (4.84) is used for coding

$$F = \{q_1, q_2, x_1, x_2, -z, \sin(z), \cos(z), \exp(z), z_1 + z_1, z_1 z_2\}.$$

Let a set of vectors of variations be given

$$W = \left(\begin{bmatrix} 5 \\ 1 \\ 6 \end{bmatrix}, \begin{bmatrix} 3 \\ 5 \\ 6 \end{bmatrix}, \begin{bmatrix} 3 \\ 10 \\ 7 \end{bmatrix}, \begin{bmatrix} 4 \\ 5 \\ 1 \end{bmatrix} \right).$$

Apply variations to the code

$$w^1 \circ C = (6, 9, 10, 7, 3, 10, 2, 6, 4, 8, 5, 3).$$

$$w^2 \circ C = (6, 9, 10, 7, 6, 3, 10, 2, 6, 4, 8, 5, 3).$$

$$w^3 \circ C = (6, 9, 10, 7, 6, 3, 10, 2, 6, 7, 4, 8, 5, 3).$$

$$w^4 \circ C = (6, 9, 10, 7, 9, 1, 3, 10, 2, 6, 7, 4, 8, 5, 3).$$

This new code describes the following mathematical expression:

$$y = \sin(\cos(q_1 + x_1) q_2 \sin(\cos(x_2)) + \exp(-x_1)).$$

4.9.3 Variational Binary Complete Genetic Programming

In the binary complete genetic programming, all codes of functions with two arguments, one argument, and arguments of the mathematical expression, including the unit elements of functions with two arguments, depending on the number of levels of the binary tree, are always located in certain places. Therefore, small variations of the code do not change the number of function arguments and do not change the length of the code.

The small variation vector contains only two components [19]

$$\mathbf{w} = [w_1 \ w_2]^T, \tag{4.151}$$

where w_1 is a position number, w_2 is a new value of the code element in accordance with the selected position.

Let a binary tree have L levels. To determine which set the element number belongs to, we first define the level at which the w_1 item is located by the formula

$$2^l \le w_1 < 2^{l+1}. \tag{4.152}$$

If the level found is not the last $l < L$, then the relation is used

$$w_2 \in \begin{cases} G_1 \text{ if } w_1 \mod 2^l \le 2^{l-1} \\ F_2, \text{ otherwise} \end{cases}. \tag{4.153}$$

If $l = L$, then

$$w_2 \in \begin{cases} G_1 \text{ if } w_1 \mod 2^l \le 2^{L-1} \\ A, \text{ otherwise} \end{cases}. \tag{4.154}$$

Let $w_1 = 1$, then the level $l = 0$, $2^0 = 1 \le w_1 = 1 < 2^1 - 2$.
Then $r = w_1 \mod 2^l = 0 \le 2^{-1} = 0.5$, therefore $w_2 \in G_1$.

Consider an Example

Let the binary complete genetic programming code of the mathematical expression (4.120) be given

$$\begin{aligned} C_0 = (&1, 1, \\ &1, 1, 2, 2, \\ &1, 3, 1, 4, \ 1, 1, 2, 2, \\ &1, 2, 1, 1, 1, 1, 1, 2, \ 1, 1, 1, 2, 1, 1, 1, 1, \\ &5, 1, 5, 1, 1, 1, 1, 1, 1, 1, 1, 1, 1, 1, 1, 1, \\ &4, 6, 5, 6, 2, 6, 4, 1, 4, 6, 5, 6, 4, 6, 3, 6). \end{aligned}$$

The code corresponds to the mathematical expression

$$y_0 = (x_1^2 - x_2^2)\cos(q_1 x_1 + q_2) + x_1 x_2 \exp(-q_3 x_1).$$

The following sets of functions and arguments are used:

$$F_2 = \{f_1 = z_1 + z_2, f_2 = z_1 z_2\},$$

$$G_1 = \{g_1 = z, g_2 = -z, g_3 = \cos(z), g_4 = \exp(z), g_5 = z^2\},$$

$$A = \{a_1 = q_1, a_2 = q_2, a_3 = q_3, a_4 = x_1, a_5 = x_2, a_6 = 0, a_7 = 1\}.$$

Introduce some vectors of variations

$$W = \left(\begin{bmatrix} 7 \\ 5 \end{bmatrix}, \begin{bmatrix} 32 \\ 4 \end{bmatrix}, \begin{bmatrix} 11 \\ 2 \end{bmatrix}, \begin{bmatrix} 48 \\ 5 \end{bmatrix} \right).$$

Applying small variations returns a new BCGP code

$$\begin{aligned}
\mathbf{w}^4 \circ \mathbf{w}^3 \circ \mathbf{w}^2 \circ \mathbf{w}^1 \circ C_0 = (&1, 1, \\
&1, 1, 2, 2, \\
&5, 3, 1, 4, \ 1, 1, 2, 2, \\
&1, 2, 1, 1, 1, 1, 1, 2, \ 1, 1, 1, 2, \\
&1, 1, 1, 1, \\
&5, 4, 5, 1, 1, 1, 1, 1, 1, 1, 1, 1, \\
&1, 1, 1, 1, \ 4, 5, 5, 6, 2, 6, 4, \\
&1, 4, 6, 5, 6, 4, 6, 3, 6).
\end{aligned}$$

The resulting code matches the following mathematical expression:

$$y_1 = (x_1^2 + \exp(x_2)) \cos(q_1 x_1 + q_2) + x_1 x_2 \exp(-q_3 x_1).$$

4.9.4 Variational Cartesian Genetic Programming

The application of the principle of small variations for Cartesian genetic programming [20] is convenient and does not require changing the code length after a small variation. The code of a mathematical expression in Cartesian genetic programming is an ordered set of vectors of calls of elementary functions. The number of components in the call vector depends on the maximum number of arguments used to encode the elementary functions. If functions with no more than three arguments are used for encoding, then the function call vector has four components, one for the function number, the rest for the argument numbers. If a function with fewer arguments is called, then the extra components of the call vector are not used.

The vector of small variations for Cartesian genetic programming consists of three components

$$\mathbf{w} = [w_1 \ w_2 \ w_3]^T, \tag{4.155}$$

where w_1 is a number of the call vector, w_2 is a number of the variable component in the call vector, w_3 is a new value of the component of the call vector. If $w_2 = 1$, then the new value of the component indicates the number of any elementary function, if $w_2 > 1$, then w_3 indicates the number of the element from the set of arguments to which the results of the calculation by the previous call vectors are added.

Consider an Example

Let the CGP code of the mathematical expression be given

$$
G_0 = \left(\begin{bmatrix} 7 \\ 4 \\ 1 \\ 2 \end{bmatrix}, \begin{bmatrix} 7 \\ 5 \\ 2 \\ 3 \end{bmatrix}, \begin{bmatrix} 7 \\ 6 \\ 3 \\ 4 \end{bmatrix}, \begin{bmatrix} 5 \\ 7 \\ 5 \\ 6 \end{bmatrix}, \begin{bmatrix} 4 \\ 8 \\ 1 \\ 2 \end{bmatrix}, \begin{bmatrix} 3 \\ 9 \\ 3 \\ 4 \end{bmatrix}, \begin{bmatrix} 6 \\ 11 \\ 12 \\ 5 \end{bmatrix}, \begin{bmatrix} 7 \\ 10 \\ 13 \\ 6 \end{bmatrix} \right). \tag{4.156}
$$

This code describes the following mathematical expression:

$$
y = \exp(q_1 x_1)(\sin(q_2 x_2) + \cos(q_3 x_3)).
$$

The following sets of arguments and functions are used:

$$
F_0 = (x_1, x_2, x_3, q_1, q_2, q_3), \tag{4.157}
$$

$$
F = \{ f_1 = z, f_2 = -z, f_3 = \cos(z), f_4 = \sin(z), f_5 = \exp(z), \\ f_6 = z_1 + z_2, f_7 = z_1 z_2, f_8 = f_{3,1}(z_1, z_2, z_3) \}, \tag{4.158}
$$

where

$$
f_{3,1}(z_1, z_2, z_3) = \begin{cases} z_2, \text{ if } z_1 \le 0 \\ z_3, \text{ otherwise} \end{cases}.
$$

Let the following set of vectors of variations be given:

$$
W = \left(\begin{bmatrix} 8 \\ 1 \\ 8 \end{bmatrix}, \begin{bmatrix} 3 \\ 3 \\ 8 \end{bmatrix}, \begin{bmatrix} 6 \\ 2 \\ 10 \end{bmatrix}, \begin{bmatrix} 1 \\ 3 \\ 2 \end{bmatrix}, \begin{bmatrix} 4 \\ 1 \\ 3 \end{bmatrix} \right). \tag{4.159}
$$

Applying small variations (4.159) to the code (4.156), and the following code of the mathematical expression is obtained:

$$
w^5 \circ w^4 \circ w^3 \circ w^2 \circ w^1 \circ G_0 =
$$

$$
\left(\begin{bmatrix} 7 \\ 4 \\ 2 \\ 2 \end{bmatrix}, \begin{bmatrix} 7 \\ 5 \\ 2 \\ 3 \end{bmatrix}, \begin{bmatrix} 7 \\ 6 \\ 8 \\ 4 \end{bmatrix}, \begin{bmatrix} 3 \\ 7 \\ 5 \\ 6 \end{bmatrix}, \begin{bmatrix} 4 \\ 8 \\ 1 \\ 2 \end{bmatrix}, \begin{bmatrix} 3 \\ 10 \\ 3 \\ 4 \end{bmatrix}, \begin{bmatrix} 6 \\ 11 \\ 12 \\ 5 \end{bmatrix}, \begin{bmatrix} 8 \\ 10 \\ 13 \\ 6 \end{bmatrix} \right). \tag{4.160}
$$

Here the first vector of variations is $w^1 = [8\ 1\ 8]^T$. Therefore, we changed the first component in the eighth call vector to the function number 8. We received $g^8 = [8\ 10\ 13\ 6]^T$. From the second vector of variations $w^2 = [3\ 3\ 8]^T$ we obtained $g^3 = [7\ 6\ 8\ 4]^T$. And so on.

The code obtained after small variations corresponds to the following mathematical expression:

$$y = \begin{cases} \sin(q_2 x_2) + \cos(\exp(q_1 x_2)), \text{ if } \cos(q_1 x_2) \leq 0, \\ q_3, \text{ otherwise.} \end{cases} \qquad (4.161)$$

As seen, the application of the principle of small variations for various methods of symbolic regression is not particularly difficult, it is enough to determine the possible variations in accordance with the encoding of mathematical expressions and set the type of their description. Genetic operations on the space of vectors of small variations are carried out in a standard way.

4.10 Multilayer Symbolic Regression Methods

The natural development of symbolic regression methods, as deep learning in artificial neural networks, is the transition to `multilayer` constructions.

The emergence of multilayer neural networks was caused by the need to approximate more complex functions, in particular, with the need to approximate the XOR function. Symbolic regression methods can directly search for complex functions, nonlinear, discontinuous, etc. However, with an increase in the dimension of the problems being solved, when searching for solutions in the form of functions of large dimensions, when the construction of the symbolic regression code also increases significantly, the use of a multilayer approach can be a promising direction.

All symbolic regression methods can be multilayered. The result of the calculation in one layer is added to the set of arguments for other layers.

In multilayer symbolic regression methods, in order to avoid cyclicality, it is necessary to establish the order of the layers and control the calls of the calculation results in the layers. The main rule for avoiding cyclical calculations is to ensure the order of using the results of calculations in layers. A higher numbered layer can use the results of calculations of lower numbered layers.

Let us consider the construction of a multilayer symbolic regression method using the example of a multilayer network operator [21].

Multilayer Network Operator

Consider the NOP code with N layers.

$$\Psi = (\Psi^1, \ldots, \Psi^N). \qquad (4.162)$$

To encode a mathematical expression, there is an ordered set of arguments of a mathematical expression and a set of functions with one and two arguments.

$$F_0 = \{x_1, \ldots, x_n, q_1 \ldots, q_p\}, \qquad (4.163)$$

$$F_1 = \{f_{1,1}(z) = z, f_{1,2}(z), \ldots, f_{1,w}(z)\}, \tag{4.164}$$

$$F_2 = \{f_{2,1}(z_1, z_2), \ldots, f_{2,v}(z_1, z_2)\}. \tag{4.165}$$

Suppose that each network operator from (4.162) has a different number of source nodes and a different number of outputs. Let us define integer vectors of the corresponding dimensions.

To describe the inputs to the network operator, the vector of inputs is introduced

$$\mathbf{r}^i = [r_1^i \ldots r_{m_i}^i]^T, \tag{4.166}$$

where i is a number of the network operator, m_i is a number of source nodes of the network operator, $i = 1, \ldots, N$.

To describe the outputs of the network operator, the output vector is used

$$\mathbf{d}^i = [d_1^i \ldots d_{n_i}^i]^T, \tag{4.167}$$

where i is a number of the network operator, n_i are numbers of network operator nodes that store the results of calculations, d_j^i is a number of the node in the network operator Ψ^i that stores the results of calculations , $j = 1, \ldots, n_i$, $i = 1, \ldots, N$, the set of output nodes must include the numbers of all sink-nodes of the network operator.

The set of arguments for the network operator on each layer includes the results of calculations on the previous layers.

Let us introduce an ordered set of the results of calculations of NOP

$$Y_i = (y_1^i, \ldots, y_{n_i}^i), \tag{4.168}$$

where y_j^i is the result of calculations of the network operator Ψ^i stored in the node numbered d_j^i, $j = 1, \ldots, n_i$, $i = 1, \ldots, N$.

Assume that the numbers of the source nodes r_j^i are the first numbers of the nodes of the network operator Ψ^i. This can always be done, since the source nodes are not directly connected to each other.

The set of arguments for each network operator is different. For the first network operator, the set of arguments coincides with the set of arguments (4.163) of the encoded mathematical expression

$$F_0^1 = F_0. \tag{4.169}$$

For each next network operator, the set of arguments includes the results of calculations of the previous network operators

$$\begin{aligned}
F_0^2 &= F_0^1 \cup Y_1, \\
F_0^3 &= F_0^2 \cup Y_2, \\
&\cdots \\
F_0^N &= F_0^{N-1} \cup Y_{N-1}.
\end{aligned} \tag{4.170}$$

Consider an Example

Let sets of arguments of the mathematical expression and functions with one and two arguments be given

$$F_0 = \{x_1, x_2, x_3, q_1, q_2, q_3\}, \tag{4.171}$$

$$
\begin{aligned}
F_1 = \{ & f_{1,1}(z) = z, f_{1,2}(z) = -z, f_{1,3}(z) = \sin(z), \\
& f_{1,4}(z) = \cos(z), f_{1,5}(z) = \exp(z), f_{1,6}(z) = \arctan(z), \\
& f_{1,7}(z) = \sqrt[3]{z}, f_{1,8}(z) = \tanh(z) \}.
\end{aligned}
\tag{4.172}
$$

$$F_2 = \{f_{1,1}(z_1, z_2 = z_1 + z_2), f_{2,2}(z_1, z_2) = z_1 z_2\}. \tag{4.173}$$

$N = 4$ network operators of different sizes are given

$$
\Psi^1 = \begin{bmatrix}
0 & 0 & 0 & 0 & 1 & 0 & 0 & 6 \\
0 & 0 & 0 & 0 & 2 & 0 & 0 & 0 \\
0 & 0 & 0 & 0 & 0 & 1 & 0 & 0 \\
0 & 0 & 0 & 0 & 0 & 2 & 0 & 0 \\
0 & 0 & 0 & 0 & 2 & 0 & 3 & 0 \\
0 & 0 & 0 & 0 & 0 & 2 & 4 & 0 \\
0 & 0 & 0 & 0 & 0 & 0 & 2 & 0 \\
0 & 0 & 0 & 0 & 0 & 0 & 0 & 1
\end{bmatrix},
\tag{4.174}
$$

$$
\Psi^2 = \begin{bmatrix}
0 & 0 & 1 & 0 & 1 & 0 \\
0 & 0 & 2 & 0 & 0 & 0 \\
0 & 0 & 2 & 5 & 0 & 0 \\
0 & 0 & 0 & 1 & 1 & 0 \\
0 & 0 & 0 & 0 & 2 & 8 \\
0 & 0 & 0 & 0 & 0 & 1
\end{bmatrix},
\tag{4.175}
$$

$$
\Psi^3 = \begin{bmatrix}
0 & 0 & 0 & 0 & 1 & 0 & 0 \\
0 & 0 & 0 & 0 & 1 & 0 & 0 \\
0 & 0 & 0 & 0 & 0 & 6 & 0 \\
0 & 0 & 0 & 0 & 0 & 1 & 0 \\
0 & 0 & 0 & 0 & 2 & 1 & 0 \\
0 & 0 & 0 & 0 & 0 & 2 & 7 \\
0 & 0 & 0 & 0 & 0 & 0 & 1
\end{bmatrix},
\tag{4.176}
$$

$$
\Psi^4 = \begin{bmatrix}
0 & 0 & 0 & 0 & 1 & 0 & 0 & 0 & 0 & 0 \\
0 & 0 & 0 & 0 & 1 & 0 & 0 & 0 & 0 & 0 \\
0 & 0 & 0 & 0 & 0 & 5 & 0 & 0 & 0 & 0 \\
0 & 0 & 0 & 0 & 0 & 6 & 0 & 0 & 0 & 0 \\
0 & 0 & 0 & 0 & 1 & 0 & 4 & 0 & 0 & 0 \\
0 & 0 & 0 & 0 & 0 & 2 & 1 & 1 & 0 & 0 \\
0 & 0 & 0 & 0 & 0 & 0 & 2 & 1 & 0 & 0 \\
0 & 0 & 0 & 0 & 0 & 0 & 0 & 1 & 1 & 0 \\
0 & 0 & 0 & 0 & 0 & 0 & 0 & 0 & 1 & 1 \\
0 & 0 & 0 & 0 & 0 & 0 & 0 & 0 & 0 & 1
\end{bmatrix}.
\tag{4.177}
$$

Input and output vectors for network operators are given

$$\mathbf{r}^1 = [1\,3\,2\,4]^T, \tag{4.178}$$

$$\mathbf{r}^2 = [3\,6]^T, \tag{4.179}$$

$$\mathbf{r}^3 = [1\,2\,7\,9]^T, \tag{4.180}$$

$$\mathbf{r}^4 = [11\,12\,8\,10]^T, \tag{4.181}$$

$$\mathbf{d}^1 = [7\,8]^T, \tag{4.182}$$

$$\mathbf{d}^2 = [5\,6]^T, \tag{4.183}$$

$$\mathbf{d}^3 = [6\,7]^T, \tag{4.184}$$

$$\mathbf{d}^4 = [10]. \tag{4.185}$$

From the output vectors, the sets of arguments for each NOP are obtained

$$F_0^1 = (x_1, x_2, x_3, q_1, q_2, q_3), \tag{4.186}$$

$$F_0^2 = (x_1, x_2, x_3, q_1, q_2, q_3, y_1^1, y_2^1), \tag{4.187}$$

$$F_0^3 = (x_1, x_2, x_3, q_1, q_2, q_3, y_1^1, y_2^1, y_1^2, y_2^2), \tag{4.188}$$

$$F_0^4 = (x_1, x_2, x_3, q_1, q_2, q_3, y_1^1, y_2^1, y_1^2, y_2^2, y_1^3, y_2^3). \tag{4.189}$$

Determine now the mathematical expression by the code of the multilayer network operator.

According to the vector of inputs (4.178) the elements of the set of arguments x_1, x_2, q_1, q_2 are fed to the first network operator to the nodes 1, 2, 3, 4 respectively. As a result of calculations, the following expressions are obtained:

$$\begin{aligned} y_1^1 &= \sin(-q_1 x_1)\cos(-q_2 x_2), \\ y_2^1 &= \sin(-q_1 x_1)\cos(-q_2 x_2) + \arctan(x_1). \end{aligned} \tag{4.190}$$

According to the vector of inputs (4.179) for the second network operator, x_3, q_3 are supplied to its input. Then

$$\begin{aligned} y_1^2 &= x_3 \exp(-q_3 x_3), \\ y_2^2 &= \tanh(x_3 \exp(-q_3 x_3)). \end{aligned} \tag{4.191}$$

According to the vector of inputs (4.180) for the third network operator, x_1, x_2, y_1^1 and y_1^2 are fed to its input. Then

$$\begin{aligned} y_1^3 &= x_1 x_2 y_1^2 \arctan(y_1^1), \\ y_2^3 &= \sqrt[3]{x_1 x_2 y_1^2 \arctan(y_1^1)}. \end{aligned} \tag{4.192}$$

According to the input vector (4.181), y_1^3, y_2^3, y_2^1, y_2^2 are fed to the input of the fourth network operator. As a result of calculations according to the matrix of the network operator, we obtain the following mathematical expression:

$$y^4 = \cos(y_1^3 + y_2^3)\exp(y_2^1)\arctan(y_2^2). \tag{4.193}$$

As a result, the mathematical expression encoded by the four-layer network operator has the following form:

$$\begin{aligned}
y = {} & \cos(x_1 x_2 y_1^2 \arctan(y_1^1)) + \\
& \sqrt[3]{x_1 x_2 y_1^2 \arctan(\sin(-q_1 x_1)\cos(-q_2 x_2)))} \times \\
& \exp(\sin(-q_1 x_1)\cos(-q_2 x_2) + \arctan(x_1)) \times \\
& \arctan(\tanh(x_3\exp(-q_3 x_3))).
\end{aligned} \tag{4.194}$$

Genetic operations for multilayer symbolic regression structures are easy to implement based on the principle of small variations. When using the principle of small variations of the basic solution for multilayer symbolic regression methods, the number of the varied layer is added in the variation code as the first component of the variation vector. The rest of the components of the vector of variations retain their previous values.

References

1. Koza, J.R.: Genetic Programming: On the Programming of Computers by Means of Natural Selection. MIT Press, Cambridge, MA/London (1992)
2. Koza, J.R., Keane, M.A., Yu, J., Bennett, F.H., Mydlowec, W., Stiffelman, O.: Automatic synthesis of both the topology and parameters for a robust controller for a non-minimal phase plant and a three-lag plant by means of genetic programming. In: Proceedings of IEEE Conference on Decision and Control, pp. 5292–5300 (1999)
3. O'Neill, M., Ryan, C.: Grammatical evolution. IEEE Trans. Evol. Comput. **5**, 349–358 (2001)
4. O'Neill, M., Ryan, C.: Grammatical Evolution. Evolutionary Automatic Programming in an Arbitrary Language. Boston, Kluwer Academic Publishers, (2003)
5. Miller, J., Thomson, P.: Cartesian genetic programming. In: Poli, R., Banzhaf, W., Langdon, W.B., Miller, J.F., Nordin, P., Fogarty, T.C. (eds.) Proceedings of EuroGP'2000R 3rd European Conference on Genetic Programming, Edinburgh, vol. 1802, pp. 121–132. Springer, Berlin (2000)
6. Miller, J.F.: Cartesian Genetic Programming. Natural Computing Series. Springer, Berlin/Heidelberg (2011)
7. Nikolaev, N., Iba, H.: Inductive genetic programming of polynomial learning networks. In: Yao, X. (ed.) Proceedings of IEEE Symposium on Combinations of Evolutionary Computation and Neural Networks ECNN-2000, pp. 158–167. IEEE-Press (2000)
8. Nikolaev, N., Iba, H.: Adaptive Learning of Polynomial Networks. Hardcover XIV. Springer, New York (2006)
9. Zelinka, I.: Analytic programming by means of SOMA algorithm. In: Proceedings of 8th International Conference on Soft Computing Mendel 02, Brno, pp. 93–101 (2002)
10. Zelinka, I., Oplatkova, Z., Nolle, L.: Analytic programming—symbolic regression by means of arbitrary evolutionary algorithms. IJSST **9**(6), 44–56 (2005)

11. Luo, C., Zhang, S.-L.: Parse-matrix evolution for symbolic regression. Eng. Appl. Artif. Intell. **25**(6), 1182–1193 (2012)
12. Diveev, A.I., Shmalko, E.Y.: Method of Binary Genetic Programming for Automation of Control Synthesis, Questions of the theory of security and stability of systems. No.19, pp. 23–39 (2017) (In Russian)
13. Diveev, A.I., Sofronova, E.A.: The network operator method for search of the most suitable mathematical equation. In: Gao, S. (ed.) Bio-inspired Computational Algorithms and Their Application, pp. 19–42. IntechOpen, Rijeka (2012)
14. Diveev, A.: A numerical method for network operator for synthesis of a control system with uncertain initial values. J. Comput. Syst. Sci. Int. **51**, 228–243 (2012)
15. Diveev, A., Shmalko, E.: Self-adjusting control for multi robot team by the network operator method. In: 2015 European Control Conference (ECC), pp. 709–714 (2015)
16. Diveev, A.I.: Small variations of basic solution method for non-numerical optimization. IFAC-PapersOnLine **48**(25), 028–033 (2015)
17. A.I. Diveev, S.I. Ibadulla, N.B. Konyrbaev, E.Yu. Shmalko, Variational genetic programming for optimal control system synthesis of mobile robots. IFAC-PapersOnLine **48**(19), 106–111 (2015); Part of special issue: 11th IFAC Symposium on Robot Control SYROCO 2015: Salvador, Brazil, 26–28 August 2015. Edited by Renato Ventura Bayan Henriques. https://doi.org/10.1016/j.ifacol.2015.12.018
18. A.I. Diveev, S.I. Ibadulla, N.B. Konyrbaev, E.Yu. Shmalko, Variational analytic programming for synthesis of optimal control for flying robot. IFAC-PapersOnLine **48**(19), 75–80 (2015); Part of special issue: 11th IFAC Symposium on Robot Control SYROCO 2015: Salvador, Brazil, 26–28 August 2015. Edited by Renato Ventura Bayan Henriques. https://doi.org/10.1016/j.ifacol.2015.12.013
19. Diveev, A., Shmalko, E.: Complete binary variational analytic programming for synthesis of control at dynamic constraints. ITM Web Conf. **10**, 02004 (2017)
20. Diveev, A., Shmalko, E.: Machine-made synthesis of stabilization system by modified cartesian genetic programming. IEEE Trans. Cybern. 1–11 (2020). https://doi.org/10.1109/TCYB.2020.3039693
21. Diveev, A., Shmalko, E.: Synthesis of control for group of autonomous robots with phase constraints by multi-layer network operator with priorities. RUDN J. Eng. Res. **18**, 115–124 (2017)

Chapter 5
Examples of MLC Problem Solutions

Abstract This chapter contains various applied examples of solving machine learning control problems by various methods of symbolic regression presented in the book. First, the tasks of unsupervised learning are considered based on the value of the target functional. The classical Pontryagin problem is considered and a comparison of the solution obtained by machine learning with the classical result is given. The problem of stabilization system synthesis for various objects is considered. Various symbolic regression methods are demonstrated. An example of solving a supervised machine learning synthesis problem is considered, where, to obtain a training sample, the optimal control problem is solved many times under different initial conditions, and then the obtained solutions are approximated by symbolic regression. An identification example is presented. An example of solving the problem of synthesized optimal control for a mobile robot in comparison with the solution of optimal control and subsequent stabilization is given. All the examples presented are aimed to show the possibilities and prospects of symbolic regression methods in machine learning control.

5.1 Control Synthesis as Unsupervised MLC

The control synthesis problem is the most important task in the field of control, and in connection with the pandemic of robotics, today it is becoming the most important task of mankind.

The use of symbolic regression methods, first of all, from our point of view, should be aimed at creating numerical methods for solving the control synthesis problem. The complexity of the mathematical expressions that are obtained as a result of solving the control synthesis problem using symbolic regression methods is not significant. On board of the control object, instead of a complex mathematical expression, you can always write its code, which can be used to calculate control values.

A. Diveev, E. Shmalko, *Machine Learning Control by Symbolic Regression*,
https://doi.org/10.1007/978-3-030-83213-1_5

105

An important role here is played by the properties that the control system acquires due to the nonlinear feedback function obtained as a result of solving the control synthesis problem using symbolic regression methods. Note that an ideal control system from the developer's point of view should be described by such a system of differential equations so that its particular solution from a given initial state always reaches the terminal state with the optimal value of the quality criterion and at the same time possesses the attractor property, i.e. attracted all the private solutions closest to it. This attractor property makes it possible to compensate for the inaccuracies of the mathematical model, measurement errors, and external disturbances. The creation of systems for stabilizing the movement of an object along an optimal trajectory is an attempt to impart the property of an attractor to some particular solution of a system of differential equations describing the dynamics of a closed-loop control system. Note that only solutions of nonlinear systems of differential equations have the properties of an attractor. In linear systems, only a stable equilibrium point can be an attractor.

Consider several examples of solving the control synthesis problem as unsupervised learning. The unsupervised machine learning for control synthesis is assumed as a direct search of the control function on the basis of the quality criterion minimization.

5.1.1 Pontryagin's Example

In the classical monograph on optimal control [1], there is an example of general control synthesis for an object described by a system of second-order linear differential equations. For comparison with the obtained solution, let us solve the same problem using the symbolic regression.

A mathematical model of the control object is given

$$
\begin{aligned}
\dot{x}_1 &= x_2, \\
\dot{x}_2 &= u.
\end{aligned}
\tag{5.1}
$$

Restrictions on controls are set

$$
-1 \le u \le 1, \, u \in U = [-1; 1].
\tag{5.2}
$$

Given the terminal state

$$
\mathbf{x}^f = [x_1^f \ x_2^f]^T = [0 \ 0]^T.
\tag{5.3}
$$

It is necessary to find a control that moves the object (5.1) from any point of the initial conditions area

$$
-2 \le x_1 \le 2, \, -1.5 \le x_2 \le 1.5, \, X_0 = [-2; 2] \times [-1.5; 1.5],
\tag{5.4}
$$

to the terminal point (5.3) in the minimum time

$$J = t_f \to \min. \tag{5.5}$$

Here the region of initial conditions is bounded (5.4) for computational reasons, and it is not the entire state space $\mathbb{R}^2$ in contrast to the classical problem formulation.

To solve the stated problem numerically by symbolic regression, instead of the region of initial conditions, a finite set of initial conditions of twenty points is set

$$
\begin{aligned}
\tilde{X}_0 = \{ & \mathbf{x}^{0,1} = [-2 \ -1.5]^T, \mathbf{x}^{0,2} = [-2 \ -0.5]^T, \mathbf{x}^{0,3} = [-2 \ 0.5]^T, \\
& \mathbf{x}^{0,4} = [-2 \ 1.5]^T, \mathbf{x}^{0,5} = [-1 \ -1.5]^T, \mathbf{x}^{0,6} = [-1 \ -0.5]^T, \\
& \mathbf{x}^{0,7} = [-1 \ 0.5]^T, \mathbf{x}^{0,8} = [-1 \ 1.5]^T, \mathbf{x}^{0,9} = [0 \ -1.5]^T, \\
& \mathbf{x}^{0,10} = [0 \ -0.5]^T, \mathbf{x}^{0,11} = [0 \ 0.5]^T, \mathbf{x}^{0,12} = [0 \ 1.5]^T, \\
& \mathbf{x}^{0,13} = [1 \ -1.5]^T, \mathbf{x}^{0,14} = [1 \ -0.5]^T, \mathbf{x}^{0,15} = [1 \ 0.5]^T, \\
& \mathbf{x}^{0,16} = [1 \ 1.5]^T, \mathbf{x}^{0,17} = [2 \ -1.5]^T, \mathbf{x}^{0,18} = [2 \ -0.5]^T, \\
& \mathbf{x}^{0,19} = [2 \ 0.5]^T, \mathbf{x}^{0,20} = [2 \ 1.5]^T \}.
\end{aligned}
\tag{5.6}
$$

The control is searched in the form of a function of coordinates of the state space

$$u = h(x_1^f - x_1, x_2^f - x_2) \in U. \tag{5.7}$$

Redefine the functional (5.5) taking into account the set of points of initial conditions (5.6)

$$\tilde{J} = \sum_{i=1}^{20} \left(t_{f,i} + p_1 \| \mathbf{x}^f - \mathbf{x}(t_{f,i} \mathbf{x}^{0,i}) \|_2 \right) \to \min, \tag{5.8}$$

where

$$t_{f,i} = \begin{cases} t, \text{ if } \| \mathbf{x}(t, \mathbf{x}^{0,i}) - \mathbf{x}^f \|_2 \le \varepsilon_0 \\ t^+, \text{ otherwise} \end{cases}, \tag{5.9}$$

$\mathbf{x}(t, \mathbf{x}^{0,i})$ is a particular solution of the system (5.1) with control (5.7) from the initial conditions $\mathbf{x}^{0,i}$, $i \in \{1, \dots, 20\}$, $t^+ = 5.1$ s, $\varepsilon_0 = 0.01$, p_1 is a weight coefficient, $p_1 = 1$,

$$\| \mathbf{x}(t, \mathbf{x}^{0,i}) - \mathbf{x}^f \|_2 = \sqrt{\sum_{i=1}^{2} (x_i(t, \mathbf{x}^{0,i}) - x_i^f)^2}. \tag{5.10}$$

To solve the problem, the method of the network operator was applied. The dimension of the NOP matrix was 14×14. As a basic solution, a limited proportional controller was chosen.

$$u = \begin{cases} 1, \text{ if } q_1(x_1^f - x_1) + q_2(x_2^f - x_2) \ge 1 \\ -1, \text{ if } q_1(x_1^f - x_1) + q_2(x_2^f - x_2) \le -1 \\ q_1(x_1^f - x_1) + q_2(x_2^f - x_2), \text{ otherwise} \end{cases}, \tag{5.11}$$

where $q_1 = 1$, $q_2 = 1$.

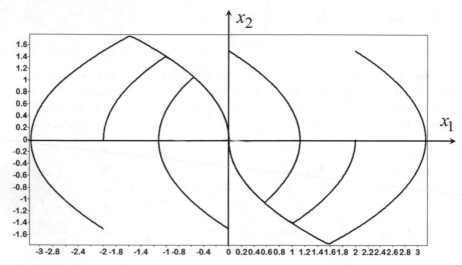

Fig. 5.1 Trajectories from eight initial conditions for control function (5.12)

As a result, the following solution was obtained:

$$u = \begin{cases} 1, \text{ if } \tilde{u} \geq 1 \\ -1, \text{ if } \tilde{u} \leq -1 \ , \\ \tilde{u}, \text{ otherwise} \end{cases} \tag{5.12}$$

where

$$\tilde{u} = \arctan(A) + \frac{1}{B} + C^3, \tag{5.13}$$

$$A = B + \arctan(q_1 \operatorname{sgn}(x_1^f - x_1)\sqrt{|x_1^f - x_1|}),$$

$$B = q_1(C + \operatorname{sgn}(C)\exp(-|C|)),$$

$$C = q_2(x_2^f - x_2) + \operatorname{sgn}(x_1^f - x_1)\sqrt{|x_1^f - x_1|}$$

$q_1 = 11.33423, q_2 = 8.13892.$

Figure 5.1 shows the simulation results of the system (5.1) with the control function (5.12) synthesized by the network operator. Simulation is performed from eight initial conditions $\mathbf{x}^{0,1} = [-2 \ -1.5]^T, \mathbf{x}^{0,4} = [-2 \ 1.5]^T, \mathbf{x}^{0,9} = [0 \ -1.5]^T,$ $\mathbf{x}^{0,12} = [0 \ 1.5]^T, \mathbf{x}^{0,17} = [2 \ -1.5]^T, \mathbf{x}^{0,20} = [2 \ 1.5]^T, \mathbf{x}^{0,21} = [-2 \ 0]^T, \mathbf{x}^{0,22} = [2 \ 0]^T.$

The obtained results can be compared with the optimal solution presented in Fig. 5.2 that shows the results of simulation from the same initial conditions with optimal control obtained using the Pontryagin maximum principle.

All trajectories in Figs. 5.1 and 5.2 coincide with high accuracy.

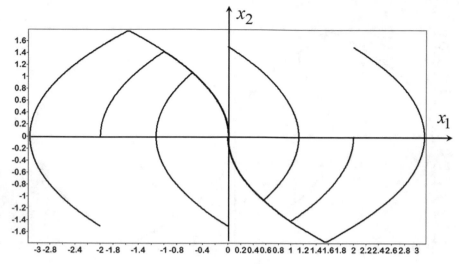

Fig. 5.2 Optimal trajectories from eight initial conditions

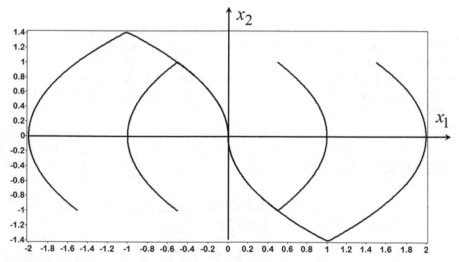

Fig. 5.3 Trajectories from eight initial conditions not from (5.6) for the control function (5.12)

Figure 5.3 shows the simulation results of the system (5.1) with the obtained control (5.12) from initial conditions that were not included in the set (5.6), which was used to solve the control synthesis problem. The simulation was carried out for the following initial conditions $\mathbf{x}^{0,23} = [-1.5\ -1]^T$, $\mathbf{x}^{0,24} = [-0.5\ -1]^T$, $\mathbf{x}^{0,25} = [0.5\ -1]^T$, $\mathbf{x}^{0,26} = [1.5\ -1]^T$, $\mathbf{x}^{0,27} = [-1.5\ 1]^T$, $\mathbf{x}^{0,28} = [-0.5\ 1]^T$, $\mathbf{x}^{0,29} = [0.5\ 1]^T$, $\mathbf{x}^{0,30} = [1.5\ 1]^T$.

For comparison, Fig. 5.4 shows the simulation results from the same initial conditions with optimal control according to Pontryagin.

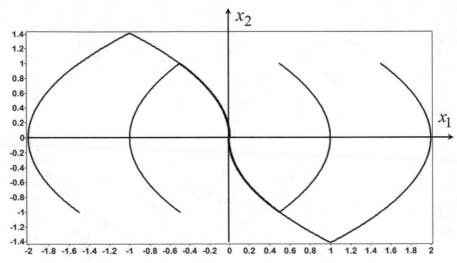

Fig. 5.4 Optimal trajectories from eight initial conditions not from set (5.6)

As can be seen from the figures, the optimal trajectories and trajectories obtained using the network operator synthesized control coincide with high accuracy. The values of the functional for all initial conditions for the optimal and synthesized controls coincided with an accuracy of 0.01 s. It follows from the simulation results that the symbolic regression method managed to synthesize the optimal control system.

5.1.2 Mobile Robot

Consider the control synthesis problem for spatial stabilization of a mobile robot.

A mathematical model of the robot is [2]

$$\begin{aligned}
\dot{x}_1 &= 0.5(u_1 + u_2)\cos(x_3), \\
\dot{x}_2 &= 0.5(u_1 + u_2)\sin(x_3), \\
\dot{x}_3 &= 0.5(u_1 - u_2),
\end{aligned} \tag{5.14}$$

where $\mathbf{x} = [x_1 \ x_2 \ x_3]^T$ is the state vector, $\mathbf{u} = [u_1 \ u_2]^T$ is the control vector.

The control values are constrained

$$-10 \le u_i \le 10, i = 1,2, \ \mathbf{u} \in U = [-10; 10] \times [-10; 10]. \tag{5.15}$$

For the model (5.14), the terminal condition is given

$$\mathbf{x}^f = [0\ 0\ 0]^T. \tag{5.16}$$

It is necessary to find the control in the form of a function of the coordinates of the state space

$$\mathbf{u} = \mathbf{h}(\mathbf{x}^f - \mathbf{x}). \qquad (5.17)$$

The function (5.17) must ensure that the robot moves to the terminal point from any initial condition of the area

$$X_0 = [-2; 2] \times [-1.5; 1.5] \times [-5\pi/12; 5\pi/12] \in \mathbb{R}^3, \qquad (5.18)$$

with time-optimal criterion

$$J = t_f \to \min. \qquad (5.19)$$

To solve the problem numerically by symbolic regression, the range of initial values is replaced with a finite set of thirty points of initial conditions

$$
\begin{aligned}
\tilde{X}_0 = \{ & \mathbf{x}^{0,1} = [-2 \ -2.5 \ -5\pi/12]^T, \mathbf{x}^{0,2} = [-2 \ -2.5 \ 0]^T, \\
& \mathbf{x}^{0,3} = [-2 \ -2.5 \ 5\pi/12]^T, \mathbf{x}^{0,4} = [-2 \ 2.5 \ -5\pi/12]^T, \\
& \mathbf{x}^{0,5} = [-2 \ 2.5 \ 0]^T, \mathbf{x}^{0,6} = [-2 \ 2.5 \ 5\pi/12]^T, \\
& \mathbf{x}^{0,7} = [-1 \ -2.5 \ -5\pi/12]^T, \mathbf{x}^{0,8} = [-1 \ -2.5 \ 0]^T, \\
& \mathbf{x}^{0,9} = [-1 \ -2.5 \ 5\pi/12]^T, \mathbf{x}^{0,10} = [-1 \ 2.5 \ -5\pi/12]^T, \\
& \mathbf{x}^{0,11} = [-1 \ 2.5 \ 0]^T, \mathbf{x}^{0,12} = [-1 \ 2.5 \ 5\pi/12]^T, \\
& \mathbf{x}^{0,13} = [0 \ -2.5 \ -5\pi/12]^T, \mathbf{x}^{0,14} = [0 \ -2.5 \ 0]^T, \\
& \mathbf{x}^{0,15} = [0 \ -2.5 \ 5\pi/12]^T, \mathbf{x}^{0,16} = [0 \ 2.5 \ -5\pi/12]^T, \\
& \mathbf{x}^{0,17} = [0 \ 2.5 \ 0]^T, \mathbf{x}^{0,18} = [0 \ 2.5 \ 5\pi/12]^T, \\
& \mathbf{x}^{0,19} = [1 \ -2.5 \ -5\pi/12]^T, \mathbf{x}^{0,20} = [1 \ -2.5 \ 0]^T, \\
& \mathbf{x}^{0,21} = [1 \ -2.5 \ 5\pi/12]^T, \mathbf{x}^{0,22} = [1 \ 2.5 \ -5\pi/12]^T, \\
& \mathbf{x}^{0,23} = [1 \ 2.5 \ 0]^T, \mathbf{x}^{0,24} = [1 \ 2.5 \ 5\pi/12]^T, \\
& \mathbf{x}^{0,25} = [2 \ -2.5 \ -5\pi/12]^T, \mathbf{x}^{0,26} = [2 \ -2.5 \ 0]^T, \\
& \mathbf{x}^{0,27} = [2 \ -2.5 \ 5\pi/12]^T, \mathbf{x}^{0,28} = [2 \ 2.5 \ -5\pi/12]^T, \\
& \mathbf{x}^{0,29} = [2 \ 2.5 \ 0]^T, \mathbf{x}^{0,30} = [2 \ 2.5 \ 5\pi/12]^T \}.
\end{aligned}
\qquad (5.20)
$$

Replace the functional

$$\tilde{J} = \sum_{i=1}^{30} \left(t_{f,i} + p_1 \| \mathbf{x}^f - \mathbf{x}(t_{f,i} \mathbf{x}^{0,i}) \|_2 \right) \to \min, \qquad (5.21)$$

where

$$t_{f,i} = \begin{cases} t, \text{ if } \| \mathbf{x}(t, \mathbf{x}^{0,i}) - \mathbf{x}^f \|_2 \leq \varepsilon_0 \\ t^+, \text{ otherwise} \end{cases}, \qquad (5.22)$$

$\mathbf{x}(t, \mathbf{x}^{0,i})$ is a particular solution of the system (5.14) with control (5.17) from the initial conditions $\mathbf{x}^{0,i}$, $i \in \{1, \ldots, 30\}$, $t^+ = 1.5$ s, $\varepsilon_0 = 0.01$, $p_1 = 1$.

To solve this machine learning control problem, the method of variational Cartesian genetic programming was used. To find a solution, a code of 20 call vectors was defined, which encoded function calls with one, two, and three arguments, so each call vector consisted of 4 components.

The proportional controller was used as a basic solution

$$u_i = \sum_{j=1}^{3} q_j(x_i^f - x_i),\qquad(5.23)$$

where $q_j = 1$, $j = 1, 2, 3$.

As a result, the following control function was obtained:

$$u_i = \begin{cases} 10, & \text{if } \tilde{u}_i \geq 10 \\ -10, & \text{if } \tilde{u}_i \leq -10 \ , \ i = 1, 2, \\ \tilde{u}, & \text{otherwise} \end{cases}\qquad(5.24)$$

where

$$\tilde{u}_1 = \mathrm{sgn}\left(-q_1^2(x_3^f - x_3)\sqrt[3]{x_2^f - x_2}\right)\sqrt{\left|-q_1^2(x_3^f - x_3)\sqrt[3]{x_2^f - x_2}\right|},\qquad(5.25)$$

$$\tilde{u}_2 = \begin{cases} q_3, & \text{if } A > B \\ -q_3, & \text{otherwise} \end{cases},\qquad(5.26)$$

$$A = x_1^f - x_1 + (q_2 + f_{3,1}((x_2^f - x_2)^{-1}, x_2^f - x_2, q_1)) \times \\ \tanh(-q_1^2(x_3^f - x_3)\sqrt[3]{x_2^f - x_2}),$$

$$B = \tanh(-q_1^2(x_3^f - x_3)\sqrt[3]{x_2^f - x_2}),$$

$$f_{3,1}(z_1, z_2, z_3) = \begin{cases} z_2, & \text{if } z_1 > 0 \\ z_3, & \text{otherwise} \end{cases},$$

$q_1 = 15.8576$, $q_2 = 10.7705$, $q_3 = 15.7634$.

The solution was obtained on the 2.8 GHz computer, Intel Core i7. To obtain the solution (5.24), the control object model with the control function was integrated more than 9,000,000 times.

Figure 5.5 shows the simulation results of the control object (5.14) with the found control function (5.24) from eight initial conditions.

$$\bar{X}_0(8). = \{[-2.5 \ -2.5 \ -5\pi/12]^T, [-2.5 \ -2.5 \ 5\pi/12]^T, \\ [-2.5 \ 2.5 \ -5\pi/12]^T, [-2.5 \ 2.5 \ 5\pi/12]^T, \\ [2.5 \ -2.5 \ -5\pi/12]^T, [2.5 \ -2.5 \ 5\pi/12]^T, \\ [2.5 \ 2.5 \ -5\pi/12]^T, [2.5 \ 2.5 \ 5\pi/12]^T\}.\qquad(5.27)$$

As you can see from the Fig. 5.5, the object moved to the terminal state not symmetrically about the x_1 axis. During the research, it was found that the most difficult points to reach the terminal position were the points located on the x_2 axis. In some cases, solutions were found with a sufficiently good value of the functional, but the obtained controls ensured the achievement of the terminal state, ignoring the initial conditions located on the x_2 axis.

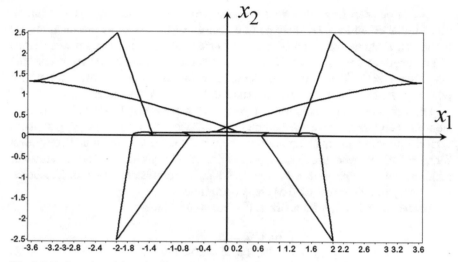

Fig. 5.5 Trajectories of the robot from 8 initial conditions with control function (5.24)

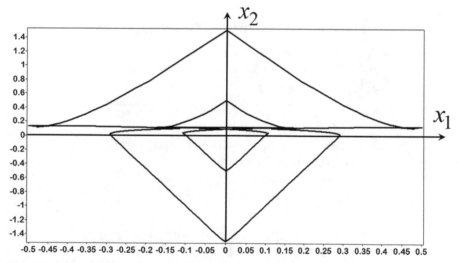

Fig. 5.6 Trajectories from 8 initial conditions not from (5.20) with control function (5.24)

To check these conditions, the system (5.14) with control (5.24) was simulated from the initial conditions located on the x_2 axis and not included in the set of initial conditions (5.20). The following initial conditions were considered $\mathbf{x}^{0,31} = [0 - 1.5 - 5\pi/12]^T$, $\mathbf{x}^{0,32} = [0 - 1.5 \, 5\pi/12]^T$, $\mathbf{x}^{0,33} = [0 - 0.5 - 5\pi/12]^T$, $\mathbf{x}^{0,34} = [0 - 0.5 \, 5\pi/12]^T$, $\mathbf{x}^{0,35} = [0 \, 1.5 - 5\pi/12]^T$, $\mathbf{x}^{0,36} = [0 \, 1.5 \, 5\pi/12]^T$, $\mathbf{x}^{0,37} = [0 \, 0.5 - 5\pi/12]^T$, $\mathbf{x}^{0,38} = [0 \, 0.5 \, 5\pi/12]^T$.

The simulation results are shown in Fig. 5.6.

As can be seen from the figure, the object reaches the terminal state from all given initial conditions. By the type of trajectory of the object's movement toward the target, it should be assumed that the movement itself is not optimal in terms of the length of the trajectory. This situation is common in unsupervised machine learning. In reality, the computer does not know intuitively how to move toward the goal, so machine intelligence has determined this type of movement.

The same control problem was solved by two more methods of symbolic regression, the network operator method and the complete binary genetic programming with the principle of small variation of the basic solution. Proportional controllers for each variable were used as a basic solution in all algorithms. The operations of addition and multiplication were used as binary operations, and a set of 28 smooth elementary functions was used as unary operations.

The network operator found the following control law:

$$u_i = \begin{cases} 10, \text{ if } \tilde{u}_i > 10 \\ -10, \text{ if } \tilde{u}_i < -10 \; , \; i = 1, 2, \\ \tilde{u}_i, \text{ otherwise} \end{cases} \tag{5.28}$$

where

$$\tilde{u}_1 = A \ln(|\tanh(C) + \mu(D) + F^3 + G + \sin(q_3 x_3)|), \tag{5.29}$$

$$\tilde{u}_2 = \tilde{u}_1 + \sin(A) + \mu(A) + B^{-1} + \text{sgn}(C) \ln(|C| + 1) + \\ \arctan(D) + \tanh(E) + \mu(G + \tanh(H) + x_1) + G - G^3 + \\ q_3^{-2} q_1^{-1} x - 1_1 + \text{sgn}(x_1), \tag{5.30}$$

$$A = B^{-1} + \sqrt[3]{\tanh(C) + \mu(D) + F^3 + G + \sin(q_3 x_3)} + \\ \text{sgn}(q_3 x_3) \exp(-|q_3 x_3|),$$

$$B = \tanh(C) + \mu(D) + F^3 + 2G + \sin(q_3 x_3) + \tanh(H) + \\ x_1 - (G + \tanh(H) + x_1)^3,$$

$$C = D + G - G^3 + \text{sgn}(q_3^2 q_1 x_1) + \arctan(q_1) + \vartheta(x_3),$$

$$D = E + \sqrt[3]{F} + \text{sgn}(G + \tanh(H) + x_1) + \text{sgn}(G) \sqrt{|G|} + \\ \arctan(q_3^2 q_1 x_1),$$

$$E = F + G + \tanh(H) + x_1 + H + \text{sgn}(q_3^2 q_1 x_1) \sqrt{|q_3^2 q_1 x_1|},$$

$$F = G + \tanh(H) + x_1 + \tanh(q_3^2 q_1 x_1) + \sqrt[3]{x_1},$$

$$G = H + \text{sgn}(q_2 x_2 + \text{sgn}(x_1)) * \exp(-|q_2 x_2 \text{sgn}(x_1)|) + \\ \cos(q_3 x_3) + \sin(x_1),$$

$$H = q_2 x_2 + \text{sgn}(x_1) + q_3 x_3 + \tanh(q_3^2 q_1 x_1),$$

$$\mu(\alpha) = \begin{cases} \alpha, \text{ if } |\alpha| \le 1 \\ \text{sgn}(\alpha), \text{ otherwise} \end{cases},$$

$q_1 = 15.80103$, $q_2 = 14.63843$, $q_3 = 13.00757$.

Results of simulation of the system (5.14) with the control function (5.29), (5.30) from eight initial conditions (5.27) are presented in the Fig. 5.7.

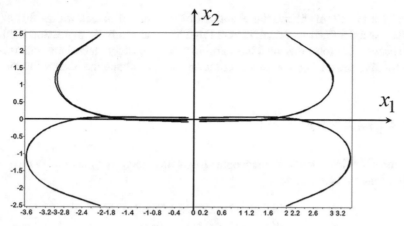

Fig. 5.7 Trajectories of the robot with control law trained by NOP

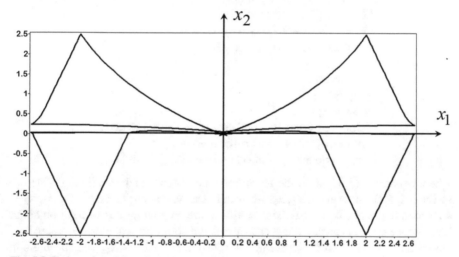

Fig. 5.8 Trajectories of the robot with control law trained by BCGP

The complete binary genetic programming found the following solution (5.28), where

$$\tilde{u}_1 = (q_2 + x_2)x_1 \cos(x_3) + \sqrt[3]{q_3} \arctan(1\sqrt[3]{x_3}) + \\ ((\sqrt[3]{q_3} + q_1)x_2 \sin(x_3)(2q_2 - q_2^3))^3,$$

(5.31)

$$\tilde{u}_2 = (2q_3 + 1)(x_2 x_1^{-1} - x_3)^{-1} + \\ (q_3^3 x_2 + \exp(1))(\sin(\exp(x_2)x_1^{-1}) - x_3),$$

(5.32)

$q_1 = 3.33594$, $q_2 = 3.77930$, $q_3 = 2.52148$.

The trajectories of the robot moving from eight initial conditions (5.27) to the terminal position (5.16) are presented in Fig. 5.8.

The goal of experiments was to show that computational symbolic regression methods allow to obtain a control function that, when substituted into the right-hand sides of a system of differential equations of the control object, makes this object stable. As a result, it was demonstrated that various symbolic regression methods can successfully solve this machine learning problem without laborious construction of a training set, basing only on the criterion for minimizing the quality functional.

5.1.3 Quadcopter

The mathematical model of quadcopter [3] is described by system of twelve ordinary differential equations

$$
\begin{aligned}
\dot{x}_1 &= x_4 + (x_5\sin(x_1) + x_6\cos(x_1))\sin(x_2)/\cos(x_2),\\
\dot{x}_2 &= (x_5\sin(x_1) + x_6\cos(x_1))/\cos(x_2),\\
\dot{x}_3 &= x_5\cos(x_1) + x_6\sin(x_1),\\
\dot{x}_4 &= x_5x_6(I_2 - I_3)/I_1 + u_1/I_1,\\
\dot{x}_5 &= x_4x_6(I_3 - I_1)/I_2 + u_2/I_2,\\
\dot{x}_6 &= x_4x_5(I_1 - I_2)/I_3 + u_3/I_3,
\end{aligned}
\tag{5.33}
$$

$$
\begin{aligned}
\dot{x}_7 &= x_{10},\\
\dot{x}_8 &= x_{11},\\
\dot{x}_9 &= x_{12},\\
\dot{x}_{10} &= u_4\sin(x_3)\cos(x_2)\cos(x_1) + \sin(x_1)\sin(x_2),\\
\dot{x}_{11} &= u_4\cos(x_3)\cos(x_1)\cos(x_2) - g,\\
\dot{x}_{12} &= u_4\cos(x_2)\sin(x_1) - \cos(x_1)\sin(x_2)\sin(x_3),
\end{aligned}
\tag{5.34}
$$

where equations (5.33) describe an angular movement, and equations (5.34) describe a spatial movement, x_1, x_3 are rotation angles about horizontal axes, x_2 is a rotation angle about vertical axis, x_4 and x_6 are angular speeds of rotation about horizontal axes, x_5 is an angular speed of rotation about vertical axis, x_7, x_9 are horizontal axes, x_8 is a vertical axis, x_{10} is a speed along axis x_7, x_{11} is a speed along axis x_8, x_{12} is a speed along axis x_9, u_i is a moment around axis x_i, $i = 1, 2, 3$, u_4 is a total lift of all four screws, g is the acceleration of gravity, I_i is a moment of inertia around axis x_i, $i = 1, 2, 3$.

To solve the synthesis problem and to achieve the object stability in the twelve-measured state space, two problems are solved consequently, the synthesis of an angular and a spatial stabilization systems.

For the first problem, the system (5.33) is used. It is necessary to find the optimal control in the following form:

$$
u_j = h_j(x_1^* - x_1, \ldots, x_6^* - x_6), \quad j = 1, \ldots, 3,
\tag{5.35}
$$

where x_i^* is a given coordinate of a point in the six-measured space $\{x_1, \ldots, x_6\}$, $i = 1, \ldots, 6$.

In the problem of angular stabilization, the following initial conditions were used:

$$
\begin{aligned}
X_0 = \{ & \mathbf{x}^{0,1} = [-0.2 \ -0.2 \ -0.2 \ 0 \ 0 \ 0]^T, \mathbf{x}^{0,2} = [-0.2 \ -0.2 \ 0 \ 0 \ 0 \ 0]^T, \\
& \mathbf{x}^{0,3} = [-0.2 \ -0.2 \ 0.2 \ 0 \ 0 \ 0]^T, \mathbf{x}^{0,4} = [-0.2 \ 0 \ -0.2 \ 0 \ 0 \ 0]^T, \\
& \mathbf{x}^{0,5} = [-0.2 \ 0 \ 0 \ 0 \ 0 \ 0]^T, \mathbf{x}^{0,6} = [-0.2 \ 0 \ 0.2 \ 0 \ 0 \ 0]^T, \\
& \mathbf{x}^{0,7} = [-0.2 \ 0.2 \ -0.2 \ 0 \ 0 \ 0]^T, \mathbf{x}^{0,8} = [-0.2 \ 0.2 \ 0 \ 0 \ 0 \ 0]^T, \\
& \mathbf{x}^{0,9} = [-0.2 \ 0.2 \ 0.2 \ 0 \ 0 \ 0]^T, \mathbf{x}^{0,10} = [0 \ -0.2 \ -0.2 \ 0 \ 0 \ 0]^T, \\
& \mathbf{x}^{0,11} = [0 \ -0.2 \ 0 \ 0 \ 0 \ 0]^T, \mathbf{x}^{0,12} = [0 \ -0.2 \ 0.2 \ 0 \ 0 \ 0]^T, \\
& \mathbf{x}^{0,13} = [0 \ 0 \ -0.2 \ 0 \ 0 \ 0]^T, \mathbf{x}^{0,14} = [0 \ 0 \ 0.2 \ 0 \ 0 \ 0]^T, \\
& \mathbf{x}^{0,15} = [0 \ 0.2 \ -0.2 \ 0 \ 0 \ 0]^T, \mathbf{x}^{0,16} = [0 \ 0.2 \ 0 \ 0 \ 0 \ 0]^T, \\
& \mathbf{x}^{0,17} = [0 \ 0.2 \ 0.2 \ 0 \ 0 \ 0]^T, \mathbf{x}^{0,18} = [0.2 \ -0.2 \ -0.2 \ 0 \ 0 \ 0]^T, \\
& \mathbf{x}^{0,19} = [0.2 \ -0.2 \ 0 \ 0 \ 0 \ 0]^T, \mathbf{x}^{0,20} = [0.2 \ -0.2 \ 0.2 \ 0 \ 0 \ 0]^T, \\
& \mathbf{x}^{0,21} = [0.2 \ 0 \ -0.2 \ 0 \ 0 \ 0]^T, \mathbf{x}^{0,22} = [0.2 \ 0 \ 0 \ 0 \ 0 \ 0]^T, \\
& \mathbf{x}^{0,23} = [0.2 \ 0 \ 0.2 \ 0 \ 0 \ 0]^T, \mathbf{x}^{0,24} = [0.2 \ 0.2 \ -0.2 \ 0 \ 0 \ 0]^T, \\
& \mathbf{x}^{0,25} = [0.2 \ 0.2 \ 0 \ 0 \ 0 \ 0]^T, \mathbf{x}^{0,26} = [0.2 \ 0.2 \ 0.2 \ 0 \ 0 \ 0]^T \}.
\end{aligned}
$$

The terminal condition was

$$
\mathbf{x}^* = [0 \ 0 \ 0 \ 0 \ 0 \ 0]^T.
$$

Restrictions on control were

$$
-2 = u_i^- \le u_i \le u_i^+ = 2, \ i = 1, 2, 3.
$$

The quality criterion was

$$
J_1 = \sum_{i=1}^{26} \left(t_{f,i} + a_1 \sqrt{ \sum_{j=1}^{6} (x_j^* - x_j(t_{f,i}, \mathbf{x}^{0,i}))^2 } \right), \tag{5.36}
$$

where a_1 is a weight coefficient, $a_1 = 1$, $t_{f,i}$ is a terminal time for solution with initial condition $\mathbf{x}^{0,i}$, $t^+ = 1.5$, $\varepsilon_1 = 0.01$, $\mathbf{x}(t, \mathbf{x}^{0,i})$ is a partial solution of the system (5.33) with control (5.35) and initial conditions $\mathbf{x}^{0,i}$, $i \in \{1, \ldots, 26\}$.

When searching for the optimal solution, parameters of the model were $I_1 = 1.5$, $I_2 = 1$, $I_3 = 1.5$, $g = 9.8067$.

For solution search the network operator method was chosen with the following parameters: a dimension of the network operator matrix 32×32, a number of functions with one argument $W = 20$, a number of functions with two arguments $V = 2$, a number of possible solutions in initial set $H = 1024$, a number of generations $P = 128$, a number of possible crossovers in one generation $R = 128$, a number of generations between change of the basic solution $E = 32$, a number of small variations of the basic solution for one possible solution $l_1 = 8$, probability of mutation $p_\mu = 0.7$. The basic solution was

$$
u_j^{(0)} = \sum_{i=1}^{6} q_i (x_i^* - x_i), \ j = 1, 2, 3, \tag{5.37}
$$

where $q_i = 1$, $i - 1, \ldots, 6$.

As a result of solving the synthesis problem by the network operator method, the following solution was obtained:

$$u_i = \begin{cases} u_i^-, & \text{if } \tilde{u}_i < u_i^- \\ u_i^+, & \text{if } \tilde{u}_i > u_i^+ \\ \tilde{u}_i - \text{otherwise} \end{cases}, \quad i = 1,2,3, \tag{5.38}$$

where

$$\tilde{u}_1 = \left(q_4(x_4^* - x_4) + q_1(x_1^* - x_1) + (x_4^* - x_4)^3 + \sqrt[3]{q_1(x_1^* - x_1)}\right)^{-1} + \\ \left(q_4(x_4^* - x_4) + q_1(x_1^* - x_1) + (x_4^* - x_4)^3 + \sqrt[3]{q_1(x_1^* - x_1)}\right)^{1/3} + \\ \text{sgn}(x_6^* - x_6) \log(|q_6(x_6^* - x_6)| + 1) + q_2(x_2^* - x_2) + \\ \text{sgn}(x_4^* - x_4)\sqrt{|q_4(x_4^* - x_4)|} + q_1(x_1^* - x_1) + (q_4(x_4^* - x_4))^3, \tag{5.39}$$

$$\tilde{u}_2 = \text{sgn}(\text{sgn}(A_1 + q_3(x_3^* - x_3) + x_3^* - x_3) \times \\ (\exp(|A_1 + q_3(x_3^* - x_3) + x_3^* - x_3|) - 1)) \times \\ (|\text{sgn}(A_1 + q_3(x_3^* - x_3) + x_3^* - x_3) \times \\ (\exp(|A_1 + q_3(x_3^* - x_3) + x_3^* - x_3|) - 1)|)^{1/2} \tag{5.40}$$

$$\tilde{u}_3 = \tanh(0.5B_1) - A_1 - q_3(x_3^* - x_3) - x_3^* + x_3 + \\ \sqrt[3]{B_1} + q_6(x_6^* - x_6) + q_2(x_2^* - x_2), \tag{5.41}$$

$$q_1 = 12.224, \ q_2 = 14.197, \ q_3 = 13.611,$$
$$q_4 = 4.361, \ q_5 = 9.989, \ q_6 = 4.144,$$
$$A_1 = \text{sgn}(x_5^* - x_5) \log(|q_5(x_5^* - x_5)| + 1),$$
$$B_1 = (A_1 + q_3(x_3^* - x_3))^3 + q_6(x_6^* - x_6) + q_2(x_2^* - x_2).$$

Figures 5.9 and 5.10 show the trajectories of quadcopter movement on the vertical plane from eight initial conditions.

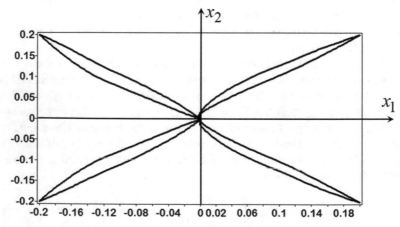

Fig. 5.9 Trajectories of the quadcopter on a plane $\{x_1, x_2\}$ from eight initial conditions

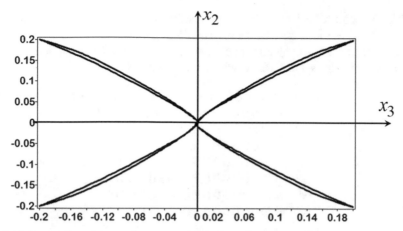

Fig. 5.10 Trajectories of the quadcopter on a plane $\{x_3, x_2\}$ from eight initial conditions

Calculation of the functional and simulations of the model was 1,035,720 times. Calculations are performed on the computer with processor Core i7, 2.8 GHz. The calculation time was about 40 minutes.

On the second stage, the full dynamic model of quad-rotor helicopter is considered. The angular model is replaced by the model with the stabilization system (5.39), (5.40), and (5.41).

$$
\begin{aligned}
\dot{x}_1 &= x_4 + (x_5 \sin(x_1) + x_6 \cos(x_1)) \sin(x_2) / \cos(x_2), \\
\dot{x}_2 &= (x_5 \sin(x_1) + x_6 \cos(x_1)) / \cos(x_2), \\
\dot{x}_3 &= x_5 \cos(x_1) + x_6 \sin(x_1), \\
\dot{x}_4 &= x_5 x_6 (I_2 - I_3)/I_1 + \tilde{h}_1(\Delta\mathbf{x})/I_1, \\
\dot{x}_5 &= x_4 x_6 (I_3 - I_1)/I_2 + \tilde{h}_2(\Delta\mathbf{x})/I_2, \\
\dot{x}_6 &= x_4 x_5 (I_1 - I_2)/I_3 + \tilde{h}_3(\Delta\mathbf{x})/I_3,
\end{aligned}
\tag{5.42}
$$

where

$$
\tilde{h}_i(\Delta\mathbf{x}) = u_i, \; i = 1, 2, 3,
\tag{5.43}
$$

$$
\Delta\mathbf{x} = \begin{bmatrix} x_1^* - x_1 \\ x_2^* - x_2 \\ x_3^* - x_3 \\ -x_4 \\ -x_5 \\ -x_6 \end{bmatrix}.
\tag{5.44}
$$

Now the optimal control system synthesis problem is solved for stabilization of object in a point of six-measure space $\{x_7, \ldots, x_{12}\}$. The control of the object includes four components

$$
\hat{\mathbf{u}} = [x_1^* \; x_2^* \; x_3^* \; u_4]^T.
\tag{5.45}
$$

So, the point $\mathbf{x}^* = [x_7^* \ldots {}_{12}^*]^T$ in six-measure space was set, and the vector of control (5.45) was searched for the systems (5.34), (5.42).

To solve this problem again the network operator method was applied.

In the problem a set of initial condition included eight elements

$$
\begin{aligned}
X_0 = \{ \mathbf{x}^{0,1} &= [0\,0\,0\,0\,0\,0 \; -0.5 \; -0.5 \; -0.5\,0\,0\,0]^T, \\
\mathbf{x}^{0,2} &= [0\,0\,0\,0\,0\,0 \; -0.5 \; -0.5\,0.5\,0\,0\,0]^T, \\
\mathbf{x}^{0,3} &= [0\,0\,0\,0\,0\,0 \; -0.5\,0.5 \; -0.5\,0\,0\,0]^T, \\
\mathbf{x}^{0,4} &= [0\,0\,0\,0\,0\,0 \; -0.5\,0.5\,0.5\,0\,0\,0]^T, \\
\mathbf{x}^{0,5} &= [0\,0\,0\,0\,0\,0\,0.5 \; -0.5 \; -0.5\,0\,0\,0]^T, \\
\mathbf{x}^{0,6} &= [0\,0\,0\,0\,0\,0\,0.5 \; -0.5\,0.5\,0\,0\,0]^T, \\
\mathbf{x}^{0,7} &= [0\,0\,0\,0\,0\,0\,0.5\,0.5 \; -0.5\,0\,0\,0]^T, \\
\mathbf{x}^{0,8} &= [0\,0\,0\,0\,0\,0\,0.5\,0.5\,0.5\,0\,0\,0]^T \}.
\end{aligned}
\tag{5.46}
$$

Constraints on control were

$$
\begin{aligned}
-\pi/4 = x_1^- &\le x_1^* \le x_1^+ = \pi/4, \\
-\pi/4 = x_2^- &\le x_2^* \le x_2^+ = \pi/4, \\
-\pi/4 = x_3^- &\le x_3^* \le x_3^+ = \pi/4, \\
0 = u_4^- &\le u_4 \le u_4^+ = 12.
\end{aligned}
\tag{5.47}
$$

Terminal condition was

$$
\mathbf{x}^f = [0\,0\,0\,0\,0\,0\,2\,0\,0\,0\,0]^T.
\tag{5.48}
$$

The following functional was used:

$$
J_2 = \sum_{i=1}^{8} \left(t_{f,i} + a_2 \sqrt{ \sum_{j=1}^{12} (x_j^f - x_j(t_{f,i}, \mathbf{x}^{0,i}))^2 } \right),
\tag{5.49}
$$

where a_2 is a weight coefficient, $a_2 = 2.5$, $t_{f,i}$ is a terminal time for solution with initial condition $\mathbf{x}^{0,i}$, $\varepsilon_1 = 0.05$, $t^+ = 2$ $\mathbf{x}(t, \mathbf{x}^{0,i})$ is a partial solution of the system (5.33) with control (5.35) and initial conditions $\mathbf{x}^{0,i}$, $i \in \{1, \ldots, 8\}$.

The network operator method had the following parameters: a dimension of the network operator matrix 32×32, a number of function with one argument $W = 16$, a number of function with two arguments $V = 2$, a number of possible solution in initial set $H = 1024$, a number of generation $P = 128$, a number of possible crossovers in one generation $R = 128$, a number of generation between change of the basic solution $E = 16$, a number small variation of basic solution for one possible solution $l_1 = 6$, probability of mutation $p_\mu = 0.7$.

A basic solution was

$$
\begin{aligned}
x_1^* &= q_{11}(x_9^f - x_9) - q_{12}, \\
x_2^* &= q_7(x_7^f - x_7) - q_8 x_{10} + q_9(x_8^f - x_8) - q_{10} x_{11} + \\
&\quad q_{11}(x_9^f - x_9) - q_{12} x_{12}, \\
x_3^* &= q_7(x_7^f - x_7) - q_8 x_{10}, \\
u_4 &= q_9(x_8^f - x_8) - q_{10} x_{11},
\end{aligned}
\tag{5.50}
$$

where $q_i = 1$, $i = 1, \ldots, 12$.

As the result, the following solution was received:

$$
x_i^* = \begin{cases} x_i^+, & \text{if } \tilde{x}_i^* > x_i^+ \\ x_i^-, & \text{if } \tilde{x}_i^* < x_i^- \\ \tilde{x}_i^* - \text{otherwise} \end{cases}, \quad i = 1, 2, 3,
\tag{5.51}
$$

$$
u_4 = \begin{cases} u_4^+, & \text{if } \tilde{u}_4 > u_4^+ \\ u_4^-, & \text{if } \tilde{u}_4 < u_4^- \\ \tilde{u}_4 - \text{otherwise} \end{cases},
\tag{5.52}
$$

$$
\tilde{x}_1^* = (A_2 q_9(x_9^f - x_9) \cos(x_{11}) \exp(-q_{12})) \sqrt[3]{A_2} \times \\
\log(|q_9(x_9^f - x_9) \cos(x_{11})|),
\tag{5.53}
$$

$$
\begin{aligned}
\tilde{x}_2^* &= \sqrt[3]{\tilde{x}_1^*} + 2 \arctan(C_2) - q_{10}^3 x_{10}^9 + D_2 - q_{11} x_{11} q_8^2 \times \\
&\quad (x_8^f - x_8)^2 - q_7(x_7^f - x_7) + \arctan(-q_{10} x_{10}^3 + \\
&\quad \arctan(A_2 q_9(x_9^f - x_9) \cos(x_{11}) \exp(-q_{12})) + \\
&\quad q_7(x_7^f - x_7)) + q_8(x_8^f - x_8) + \\
&\quad \operatorname{sgn}(-q_{10} x_{10}^3 + q_7(x_7^f - x_7)) \sqrt{|-q_{10} x_{10}^3 + q_7(x_7^f - x_7)|} \\
&\quad + \tanh(-0.5 x_{12}) + \mu (A_2 q_9(x_9^f - x_9) \cos(x_{11}) \exp(-q_{12})) \\
&\quad + \operatorname{sgn}(E_2) * (\exp(|E_2|) - 1),
\end{aligned}
\tag{5.54}
$$

$$
\tilde{x}_3^* = \operatorname{sgn}(\tilde{x}_1^*) \log(|\tilde{x}_1^*| + 1) + \sqrt[3]{B_2} + \arctan(C_2) - q_{10}^3 x_{10}^9 - \\
\tilde{x}_1^* + (-q_{10} x_{10}^3 + q_7(x_7^f - x_7))^3,
\tag{5.55}
$$

$$
\begin{aligned}
\tilde{u}_4 &= \sin(\tilde{x}_3^*) + \operatorname{sgn}(\tilde{x}_2^*) * (\exp(|\tilde{x}_2^*|) - 1) + \\
&\quad \operatorname{sgn}(\tilde{x}_1^*) \log(|\tilde{x}_1^*| + 1) + C_2^2 + (A_2 q_9(x_9^f - x_9) \cos(x_{11}) \times \\
&\quad \exp(-q_{12}))^2 + \tanh(0.5 B_2) + \tanh(0.5 E_2) + \\
&\quad (-q_{10} x_{10}^3 + q_7(x_7^f - x_7))^2,
\end{aligned}
\tag{5.56}
$$

where

$$
A_2 = q_{12} - x_{12} + \sqrt[3]{q_{10}} + \arctan(q_9) + \cos(x_7^f - x_7),
$$

$$
\begin{aligned}
B_2 &= \arctan(-q_{10} x_{10}^3 + q_7(x_7^f - x_7)) - q_{11} x_{11} q_8^2 (x_8^f - x_8)^2 + \\
&\quad q_8(x_8^f - x_8) - q_7(x_7^f - x_7),
\end{aligned}
$$

$$C_2 = \text{sgn}(-q_{10}x_{10}^3 + q_7(x_7^f - x_7)) * (\exp(|-q_{10}x_{10}^3 + q_7(x_7^f - x_7)|) - 1) + \text{sgn}(-q_{10}x_{10}^3)\sqrt{|q_{10}x_{10}^3|} + \mu(q_7(x_7^f - x_7)) + q_7^3,$$

$$D_2 = 2(\arctan(-q_{10}x_{10}^3 + q_7(x_7^f - x_7)) - q_{11}x_{11}q_8^2(x_8^f - x_8)^2 + \text{sgn}(-q_{11}x_{11}q_8^2(x_8^f - x_8)^2)\exp(-|-q_{11}x_{11}q_8^2(x_8^f - x_8)^2|) + q_8(x_8^f - x_8) + \exp(q_{10}) - 1 + (-q_{10}x_{10}^3 + q_7(x_7^f - x_7))^3 - q_7(x_7^f - x_7)),$$

$$E_2 = \arctan(-q_{10}x_{10}^3 + q_7(x_7^f - x_7)) - q_{11}x_{11}q_8^2(x_8^f - x_8)^2 + q_8(x_8^f - x_8) - q_7(x_7^f - x_7),$$

$$\mu(\beta) = \begin{cases} \text{sgn}(\beta), & \text{if } |\beta| > 1 \\ \beta - \text{otherwise} \end{cases},$$

$q_7 = 0.115$, $q_8 = 3.371$, $q_9 = 3,076$, $q_{10} = 0,144$, $q_{11} = 3.131$, $q_{12} = 4.515$.

At the search process, the criterion (5.49) was counted 1,189,440 times. Calculation time was about 1.5 hour.

In the Figs. 5.11 and 5.12 the quadcopter trajectories from eight initial conditions are shown.

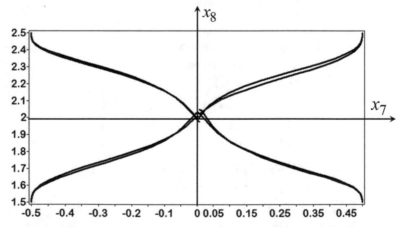

Fig. 5.11 Trajectories of quadcopter in vertical plane $\{x_7, x_8\}$ from eight initial conditions

Plots of controls in spatial spaces are shown in the Figs. 5.13, 5.14, 5.15, and 5.16.

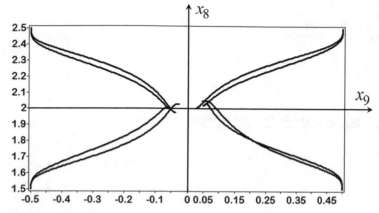

Fig. 5.12 Trajectories of quadcopter in vertical plane $\{x_9, x_8\}$ from eight initial conditions

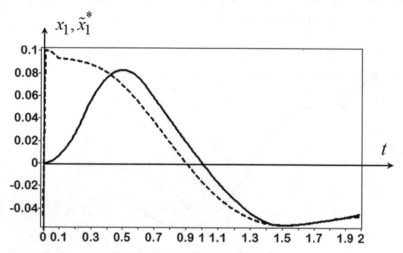

Fig. 5.13 Control $\tilde{x}_1^*$ (dash) and coordinate x_1

5.2 Control Synthesis as Supervised MLC

Supervised machine learning control is a learning with application of a training set. In this case firstly it is necessary to create a training set in order to show to the learning object what we want of it. For this purpose, initially the optimal control problem from some different initial conditions can be solved with the same quality criterion as for the synthesis problem. Obtained optimal trajectories are templates for learning. They show what forms of plots for variables must be obtained in the

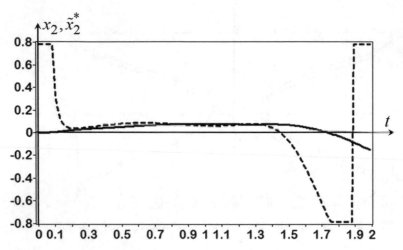

Fig. 5.14 Control $\tilde{x}_2^*$ (dash) and coordinate x_2

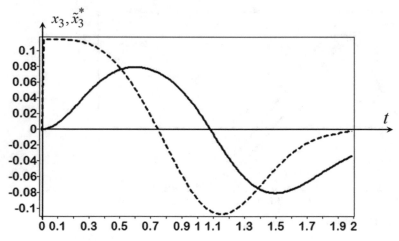

Fig. 5.15 Control $\tilde{x}_3^*$ (dash) and coordinate x_3

result of control synthesis problem solution and what values of functional must give these solutions. Then, obtained optimal trajectories for different initial conditions are approximated by some symbolic regression method.

Let us demonstrate the proposed approach of machine learning based on approximation of optimal trajectories in the computational example of general synthesis of optimal control for a spacecraft landing on the surface of the Moon [4].

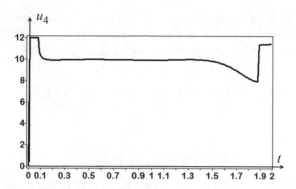

Fig. 5.16 Control u_4 of total thrust of all quadcopter screws

The mathematical model of the spacecraft landing is described by the following system of differential equations:

$$\dot{x}_1 = \frac{g_E(P_c + u_2)\cos(u_1 - x_2)}{x_5} - g_{Mh}\cos(x_2),$$

$$\dot{x}_2 = \frac{g_E(P_c + u_2)\sin(u_1 - x_2)}{x_5} + \frac{g_{Mh}\sin(x_2)}{x_1},$$

$$\dot{x}_3 = \frac{x_1\cos(x_2)}{1000}, \tag{5.57}$$

$$\dot{x}_4 = \frac{x_1\sin(x_2)}{1000},$$

$$\dot{x}_5 = -\frac{P_c + u_2}{P_s},$$

where $\mathbf{x} = [x_1 \ x_2 \ x_3 \ x_4 \ x_5]^T$ is a state vector, namely x_1 is the current speed of the spacecraft (m/s), x_2 is a trajectory inclination angle (rad), x_3 is the current flight altitude relative to the lunar surface (km), x_4 is a flight distance (km), x_5 is the mass of spacecraft including fuel (kg),
$\mathbf{u} = [u_1 \ u_2]^T$ is a control vector, values of which are constrained

$$-\frac{\pi}{2} \le u_1 \le \frac{\pi}{2}, \quad -80 \le u_2 \le 80. \tag{5.58}$$

Parameters of the model have the following values: gravitational acceleration at the certain altitude above the lunar surface

$$g_{Mh} = g_M \left(\frac{r_M}{r_M + x_3}\right)^2, \tag{5.59}$$

the Moon gravitational acceleration $g_M = 1.623\,\mathrm{m/s^2}$, the Earth gravitational acceleration $g_E = 9.80665\,\mathrm{m/s^2}$, the Moon radius $r_M = 1737\,\mathrm{km}$, nominal thrust of the spacecraft engine $P_c = 720\,\mathrm{kg}$, spacecraft engine thrust $P_s = 319\,\mathrm{s}$.

A domain of initial states is

$$X_0 = \{x_{0,1} = 1689, \ -1.65 \le x_{0,2} \le -1.55, \atop 17 \le x_{0,3} \le 20, \ x_{0,4} = 0, \ x_{0,5} = 1500\}. \tag{5.60}$$

A terminal state is

$$\mathbf{x}^f = \left[x_1^f = 10, \ x_3^f = 0.2 \right]^T. \tag{5.61}$$

Phase constraints are determined by the mechanics of spacecraft flight. Obviously, the speed x_1, altitude x_3, and fuel level x_5 cannot be negative, reaching a zero altitude x_3 or zero fuel level x_5 at a significant speed x_1 means that the spacecraft has crashed.

Consider the following phase constraints:

$$h_k(\mathbf{x}) = -x_j \le 0, \ k = 1,2,3, \ j = 1,3,5,$$

$$h_k(\mathbf{x}) = \vartheta(0.001 - x_j)(x_1 - V_{\max}) \le 0, \ k = 4,5, \ j = 3,5, \tag{5.62}$$

where $V_{\max}$ is the maximum landing speed, $V_{\max} = 1$, $\vartheta(A)$ is the Heaviside step function

$$\vartheta(A) = \begin{cases} 1, \text{ if } A > 0 \\ 0, \text{ otherwise} \end{cases}.$$

According to the proposed method at the first step, the training set is to be formed. We determine the finite set of initial states within the domain (5.60) and solve the optimal control problem for each initial state from this set.

Let us replace the domain of initial states (5.60) with a set of $M = 21$ elements uniformly distributed on this domain

$$\tilde{X}_0 = \{\mathbf{x}^{0,j+7(i-1)} = [1689 \ -1.65 + 0.05(i-1) \atop 17 + 0.5(j-1) \ 0 \ 1500]^T\}, i = \overline{1,3}, j = \overline{1,7}. \tag{5.63}$$

Quality criterion considers the proximity of reaching terminal state and violation of phase constraints

$$J_j = \alpha_1 \sqrt{\left(x_1\left(t_f(\mathbf{x}^{0,j})\right) - x_1^f \right)^2 + \left(x_3\left(t_f(\mathbf{x}^{0,j})\right) - x_3^f \right)^2} +$$

$$\alpha_2 \int_0^{t_f(\mathbf{x}^{0,j})} \left(\sum_{k=1}^K \vartheta\left(h_k(\mathbf{x}(t,\mathbf{x}^{0,j}))\right) h_k(\mathbf{x}(t,\mathbf{x}^{0,j})) \right) dt \to \min, \tag{5.64}$$

where α_i are given penalty coefficients, $i = 1,2$, $K = 5$ is a number of phase constraints, $j = \overline{1,M}$.

To search for solution to the optimal control problem the direct approach was used. The original problem was reduced to a nonlinear programming problem by introducing the time interval Δt. The solution of each optimal control problem in the form of control vector at discrete moments of time was searched independently by hybrid evolutionary algorithm combining Grey Wolf Optimizer (GWO) [7] and Particle swarm optimization (PSO) [5].

In a computational experiment the size of the set of possible solutions was 100, number of search iterations was 5000. Modeling parameters were the following: maximum control time $t_{max} = 300$, discretization time interval $\Delta t = 30$, penalty factors $\alpha_1 = 10$, $\alpha_2 = 10$.

At the second step of proposed approach, we use obtained optimal trajectories to synthesize a multidimensional control function of object state space. The search for a control function is conducted by a symbolic regression method that searches for the most suitable expression that approximates provided optimal trajectories best.

We used the network operator method to synthesize a control function. In the computational experiment we used the following parameters of NOP: size of NOP matrix was 40, size of the set of input variables was 3, size of the set of input parameters was 12, number of outputs was 2, number of candidate solutions in the initial set was 256, maximum number of search iteration was 25,000.

As a result of computational experiment, a control function in the form of NOP matrix and a parameter vector was obtained. The mathematical expression for the found control function is as follows:

$$u_1 = \begin{cases} \pi/2, & \text{if } \tilde{u}_1 > \pi/2 \\ -\pi/2, & \text{if } \tilde{u}_1 < -\pi/2 \\ \tilde{u}_1, & \text{otherwise} \end{cases}, \quad u_2 = \begin{cases} 80, & \text{if } \tilde{u}_2 > 80 \\ -80, & \text{if } \tilde{u}_2 < -80 \\ \tilde{u}_2, & \text{otherwise} \end{cases}, \quad (5.65)$$

where

$$\tilde{u}_1 = \chi_6(-z_{34}, \tanh(z_{35}), z_{37}),$$

$$\tilde{u}_2 = \operatorname{sgn}(u_1)\sqrt{|u_1|} - z_{36} + \arctan(z_{35}) + \operatorname{sgn}(z_{34})\sqrt{|z_{34}|} + \log(|z_{33}|) +$$
$$z_{32}^{-1} + z_{31} + z_{28} - z_{28}^3 + \log(|z_{26}|) - z_{25} + \arctan(z_{21}) + \operatorname{sgn}((z_{17}))\sqrt{|z_{17}|} +$$
$$\exp(q_{12}) + \tanh(q_{10}) + q_9 + \log(q_5) + \tanh(q_1),$$

$$z_{37} = \min\{z_{36} - z_{36}^3, \log(|z_{35}|), z_{34} - z_{34}^3, z_{32}^{-1}, \operatorname{sgn}(z_{28})\sqrt{|z_{28}|}, \tanh(z_{27}), \exp(z_{25}), z_{20}^{-1}, \sqrt[3]{x_2}\},$$

$$z_{36} = \min\{\exp(z_{29}), \tanh(z_{28}), \operatorname{sgn}(z_{27})\sqrt{|z_{27}|}, \arctan(z_{26}), \sqrt[3]{z_{22}}, z_{20}^3, \tanh(q_6), \sqrt[3]{q_1}, x_3^3\},$$

$$z_{35} = \arctan(z_{23}) + \tanh(z_{22}) + \tanh(q_{12}) + \log(|x_3|),$$

$$z_{34} = \max\{\operatorname{sgn}(z_{33})\sqrt{|z_{33}|}, \log(|z_{30}|), z_{29}^{-1}, z_{22}^{-1}, z_{20}^{-1}\},$$

$$z_{33} = \min\{z_{32}^{-1}, \arctan(z_{29}), \tanh(z_{24}), z_{22}^3, z_{19}, -z_{16}, z_{11}^{-1}, \tanh(q_3), \tanh(q_2), \exp(x_2)\},$$

$$z_{32} = \min\{z_{31}^3, z_{26}, \log(|z_{25}|), z_{18}^{-1}, \sqrt{q_{10}}, \arctan(q_6), \sqrt[3]{q_2}\},$$

$$z_{31} = \log(|z_{27}|) + z_{26}^2 + z_{24}^2 + \arctan(z_{22}) + \sqrt[3]{z_{21}} + \exp(z_{20}) + \exp(z_{17}) +$$
$$z_{16}^{-1} + q_{12}^3 - q_5 + \sqrt[3]{q_2} + \sqrt[3]{x_1},$$

$$z_{30} = \max\{z_{29}, -z_{26}, \tanh(z_{25}), \arctan(z_{18}), \operatorname{sgn}(z_{16})\sqrt{|z_{16}|}, q_{11}\},$$

$$z_{29} = \chi_6(z_{28}, z_{27}^{-1}, z_{26}^3, \log(|z_{24}|), -z_{22}^3, -z_{20}, \sqrt[3]{z_{17}}, q_7, q_2^{-1}, \sqrt{q_1}),$$

$$z_{28} = \max\{z_{27}, z_{23}^{-1}, -z_{20}, \log(|z_{19}|), z_{18} - z_{18}^3, -z_{17}, \log(|z_{16}|), q_7, \log(q_5)\},$$

$$z_{27} = \chi_6(z_{24}, \arctan(z_{20}), \sqrt{q_{11}}, q_8 - q_8^3, q_6^3, -q_5, \log(q_4)),$$

$$z_{26} = \chi_5(\sqrt[3]{z_{23}}, -z_{20}, z_{19} - z_{19}^3, -z_{18}, \mathrm{sgn}(z_{15})\sqrt{|z_{15}|}, q_{12}, \sqrt[3]{x_3}),$$

$$z_{25} = \max\{z_{23}^2, z_{22}, z_{21} - z_{21}^3, \arctan(z_{17}), q_{11}, q_9, \sqrt{q_5}\},$$

$$z_{24} = \max\{\tanh(z_{21}), z_{20} - z_{20}^3, z_{15}^3, q_{12} - q_{12}^3, q_{10}, q_8^3, \sqrt[3]{q_5}, q_4^3, \exp(x_2)\},$$

$$z_{23} = z_{20}z_{18}z_{16}q_9\sqrt{q_7}\,\mathrm{sgn}(x_1)\sqrt{|x_1|},$$

$$z_{22} = \chi_6(z_{19}, z_{17}^{-1}, \exp(z_{16}), \sqrt[3]{q_{10}}, \arctan(q_9), q_8^{-1}, \sqrt{q_2}),$$

$$z_{21} = \max\{\tanh(z_{18}), q_8^2, q_7, -q_6, \sqrt{q_3}\},$$

$$z_{20} = \chi_6(-z_{19}, z_{17}, \arctan(z_{10}), q_6, -q_2, x_3^{-1}),$$

$$z_{19} = z_{16} + \arctan(q_{11}) + q_8^{-1} + q_5 + \arctan(x_3),$$

$$z_{18} = \chi_5(z_{15}, q_{12}^{-1}, q_7^{-1}, q_4, x_1 - x_1^3),$$

$$z_{17} = q_3 + x_3 - q_3 x_3,$$

$$z_{16} = q_2 x_2 \tanh(q_5),$$

$$z_{15} = \mathrm{sgn}(\sqrt{q_{10}} + q_5 - q_5^3 + q_3^{-1} + q_1 + x_1)\sqrt{q_{10} + (q_5 - q^3)^2 + q_3^{-2} + q_1^2 + x_1^2},$$

$$\chi_5(a_1, a_2) = a_1 + a_2 - a_1 a_2,$$

$$\chi_5(a_1, \ldots, a_s) = \chi_5(a_1, \chi_5(a_2, \chi_5(\ldots, \chi_5(a_{s-1}, a_s)\ldots))),$$

$$\chi_6(a_1, \ldots, a_s) = \mathrm{sgn}\left(\sum_{i=1}^{s} a_i\right)\sqrt{\sum_{i=1}^{s} a_i^2},$$

$q_1 = 2.3474$, $q_2 = 10.5066$, $q_3 = 9.9106$, $q_4 = 13.1419$, $q_5 = 9.6631$, $q_6 = 4.4541$, $q_7 = 2.1899$, $q_8 = 4.8552$, $q_9 = 3.1116$, $q_{10} = 6.6172$, $q_{11} = 12.6812$, $q_{12} = 15.6148$.

Further, the obtained solution was tested for various initial conditions from (5.60), both those that were used in training (5.63) (see Table 5.1) and new ones that were not used (see Table 5.2).

Table 5.1 shows the quality of approximation. Here J^* are the values of quality criterion obtained using the found control function (5.65) for 21 initial states previously used as a training set. The value of the quality criterion J_{opt} obtained by solving the optimal control problem for the same initial state is presented as a reference value. The average deviation of the quality criterion values from the reference ones is 0.0591, maximum deviation is 0.2514, the standard deviation is 0.0648.

Table 5.2 shows the values of quality criterion J^* obtained using the found control function (5.65) for 10 initial states generated randomly within the domain (5.60). This test shows the efficiency of the found control function for any initial state from the domain (5.60). The value of the quality criterion J_{opt} obtained by solving the optimal control problem for the same initial state is shown in the table as a reference value. The average deviation of the quality criterion values from the reference ones is 0.0366, maximum deviation is 0.1122, the standard deviation is 0.0341.

Table 5.1 Results of testing using initial states from the training set

Initial state $\mathbf{x}^0$				J^*	J_{opt}
$[1689 - 1.65$	17	0	$1500]^T$	0.1777	0.0018
$[1689 - 1.6$	17	0	$1500]^T$	0.0295	0.0056
$[1689 - 1.55$	17	0	$1500]^T$	0.0049	0.0029
$[1689 - 1.65$	17.5	0	$1500]^T$	0.1433	0.0060
$[1689 - 1.6$	17.5	0	$1500]^T$	0.0264	0.0044
$[1689 - 1.55$	17.5	0	$1500]^T$	0.0127	0.0024
$[1689 - 1.65$	18	0	$1500]^T$	0.0780	0.0033
$[1689 - 1.6$	18	0	$1500]^T$	0.0439	0.0084
$[1689 - 1.55$	18	0	$1500]^T$	0.0061	0.0023
$[1689 - 1.65$	18.5	0	$1500]^T$	0.0703	0.0019
$[1689 - 1.6$	18.5	0	$1500]^T$	0.0111	0.0019
$[1689 - 1.55$	18.5	0	$1500]^T$	0.2525	0.0011
$[1689 - 1.65$	19	0	$1500]^T$	0.0240	0.0012
$[1689 - 1.6$	19	0	$1500]^T$	0.0501	0.0024
$[1689 - 1.55$	19	0	$1500]^T$	0.0030	0.0009
$[1689 - 1.65$	19.5	0	$1500]^T$	0.1035	0.0036
$[1689 - 1.6$	19.5	0	$1500]^T$	0.0822	0.0027
$[1689 - 1.55$	19.5	0	$1500]^T$	0.0045	0.0045
$[1689 - 1.65$	20	0	$1500]^T$	0.0954	0.0036
$[1689 - 1.6$	20	0	$1500]^T$	0.0635	0.0052
$[1689 - 1.55$	20	0	$1500]^T$	0.0334	0.0099

Table 5.2 Results of testing for random initial states

Initial state $\mathbf{x}^0$	J^*	J_{opt}
$[1689 - 1.565\ 18.92\ 0\ 1500]^T$	0.0176	0.0034
$[1689 - 1.571\ 17.21\ 0\ 1500]^T$	0.0048	0.0018
$[1689 - 1.63\ 19.39\ 0\ 1500]^T$	0.1136	0.0014
$[1689 - 1.558\ 19.91\ 0\ 1500]^T$	0.0567	0.0083
$[1689 - 1.582\ 17.62\ 0\ 1500]^T$	0.0042	0.0016
$[1689 - 1.628\ 18.2\ \ 0\ 1500]^T$	0.0398	0.0016
$[1689 - 1.614\ 19.24\ 0\ 1500]^T$	0.0508	0.0022
$[1689 - 1.644\ 18.57\ 0\ 1500]^T$	0.0613	0.0052
$[1689 - 1.563\ 18.33\ 0\ 1500]^T$	0.0032	0.0023
$[1689 - 1.589\ 18.9\ \ 0\ 1500]^T$	0.0464	0.0043

Figures 5.17 and 5.18 illustrate the results of computational experiment for initial state $\mathbf{x}^0 = [1689 - 1.565\ 18.92\ 0\ 1500]^T$. Trajectories obtained using the found con-

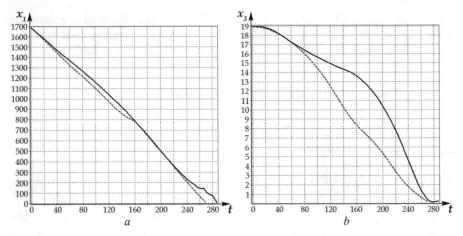

Fig. 5.17 The graphs of trajectories: (**a**) spacecraft speed over time $x_1(t)$; (**b**) spacecraft altitude over time $x_3(t)$. Found solution—black solid line; reference solution—grey dashed line

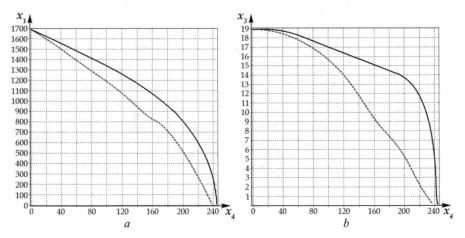

Fig. 5.18 The graphs of trajectories: (**a**) spacecraft speed over distance $x_1(x_4)$; (**b**) spacecraft altitude over distance $x_3(x_4)$. Found solution—black solid line; reference solution—grey dashed line

trol function are shown by black solid lines, and the optimal trajectories are shown by grey dashed lines. Figure 5.19 shows the found control function values over time.

The computational experiment showed that the found multidimensional control function allows to obtain a close to optimal solution for any initial states from the given domain (5.60) even for those initial states that were not in the training set (5.63).

The presented example illustrates the methodology for solving the control synthesis problem as supervised machine learning control based on a training set. The training sample is constructed based on multiple solutions of the optimal control problem. An example shows that the machine learning control, obtained by sym-

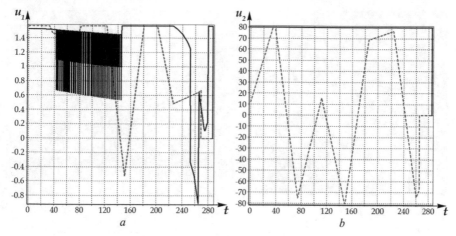

Fig. 5.19 The graphs of control values over time: (**a**) control $u_1(t)$; (**b**) control $u_2(t)$. Found solution—black solid line; reference solution—grey dashed line

bolic regression, gives good results not only for the input data from the training set but also not from it.

5.3 Identification and Control Synthesis for Multi-link Robot

In the present example, we are going to consider the solution of the problem of identifying a mathematical model of a control object using machine learning by symbolic regression methods. In this example, we not only identify the object model but also build a control system for it, solving the synthesis problem for the identified object model.

The identification problem emerges from the unknown or extremely complex nature of control objects. The derivation of the model often requires significant time or even impossible using traditional methods. And as mentioned in Chap. 2, identification also requires automation by machine learning methods.

Consider an example of identification and control synthesis [6] for multi-link robot presented in (Fig. 5.20).

Assume a model of control object is unknown.

There is the real object and for it some experiments can be conducted. It is known that the control vectors are electromagnetic moments u_i which act on the rotor R_i. Components of the state vector are the joint angles ϕ_i, $i = 1, 2, 3$.

In the experiment, to obtain the original data for the robot the following time dependent test control functions were used:

$$\begin{aligned}
u_1 &= 0.35\sin(t), \\
u_2 &= 0.3(3-t), \\
u_3 &= 0.15\sin(t) + 0.3\cos(t).
\end{aligned} \qquad (5.66)$$

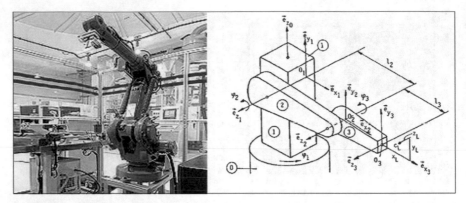

Fig. 5.20 A multi-link robot

A series of experiments was conducted, and the values **x** of variables in 21 points were obtained. Such small number of points turned out to be sufficient for such type of trajectory. Experimental data are presented in Table 5.3.

Table 5.3 Experimental Data

t (s)	x_1 (rad)	x_2 (rad)	x_3 (rad)	u_1	u_2	u_3
0	0	0	0	0	0.9	0.3
0.1	0	0	−0.32	0.03	0.87	0.313
0.2	0	−0.028	−0.102	0.07	0.84	0.323
0.3	0	−0.05	−0.250	0.1034	0.81	0.3309
0.4	−0.007	−0.084	−0.4111	0.1363	0.78	0.3347
0.5	−0.01	−0.146	−0.6418	0.1678	0.75	0.3352
0.6	−0.02	−0.221	−0.8503	0.1976	0.72	0.3323
0.7	−0.048	−0.35	−1.042	0.2255	0.69	0,1261
0.8	−0.07	−0.468	−1.223	0.2511	0.66	0.3166
0.9	−0.116	−0.64	−1.319	0.2742	0.63	0.304
1,0	−0.165	−0.79	−1.323	0.2945	0.6	0.2883
1.1	−0.246	−0.977	−1.247	0.3119	0.57	0.2689
1.2	−0.313	−1.093	−1.085	0.3262	0.54	0.2485
1.3	−0.44	−1.216	−0.8576	0.3372	0.51	0.2248
1.4	−0.514	−1.32	−0.5798	0.3449	0.48	0.1998
1.5	−0.642	−1.407	−0.2505	0.3491	0.45	0.1708
1.6	−0.754	−1.452	0.007	0.3499	0.42	0.1412
1.7	−0.89	−1.463	0.2505	0.3471	0.39	0.1101
1.8	−1.01	−1.431	0.3952	0.3408	0.36	0.0779
1.9	−1.18	−1.353	0.4444	0.3312	0.33	0.045
2.0	−1.29	−1.23	0.373	0.3183	0.3	0.0115

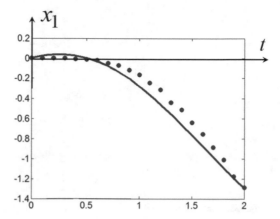

Fig. 5.21 The plot for $x_1(t)$

The identification was conducted by the network operator method. The simple redundant linear model of the control object was taken as the basic solution

$$\dot{x}_1 = q_1 x_1 + q_2 x_2 + q_3 x_3 + u_1 + q_4,$$
$$\dot{x}_1 = q_1 x_1 + q_2 x_2 + q_3 x_3 + u_2 + q_5, \qquad (5.67)$$
$$\dot{x}_1 = q_1 x_1 + q_2 x_2 + q_3 x_3 + u_3 + q_6.$$

As a result, the following solution was obtained:

$$\dot{x}_1 = u_1^3 + u_2^3 + u_3^3 + q_1 u_2^2 + u_1 q_4 \cos(u_3),$$
$$\dot{x}_2 = 2x_1 - 2x_1^3 - q_3 + q_1^3 + \cos(x_2) + x_2 - x_2^3 + u_2 + q_5,$$
$$\dot{x}_3 = x_2 - x_2^3 + 2u_1 q_4 \cos(u_3) - (u_1 q_4 \cos(u_3))^3 + u_1^3 + u_3^3 + q_1 u_2^2 - \qquad (5.68)$$
$$\quad (u_1^3 + u_3^3 + q_1 u_2^2 + u_1 q_4 \cos(u_3))^3 + \sin(\cos(x_1) \cos(x_2) \cos(x_3) \times$$
$$\quad q_3 u_2 + u - 2^3 + u_3 + q_6 + \cos(x_1 - x_1^3 + x_2 - x_2^3 + u_2 + q_5)),$$

where $q_1 = 0.234375$, $q_2 = 0.984375$, $q_3 = 3.875000$, $q_4 = 3.984375$, $q_5 = 0.484375$, $q_6 = 0.0000$.

Results of simulation of the obtained control system model are presented in Figs. 5.21, 5.22, and 5.23. Points indicate the experimental data from Table 5.3, and lines are the results of simulation.

Figures 5.21, 5.22, and 5.23 show that identified mathematical model is good enough for the experimental data, though at $t = 1$ s the error is approximately 10%.

At the second step, we validate the solution of the identification problem and solve the control synthesis problem using the model we have derived at the identification step. To solve the synthesis problem we also use the network operator method of symbolic regression.

As the control goal, we set a trajectory that should be followed by the robot from point to point (see Figs. 5.24, 5.25, and 5.26).

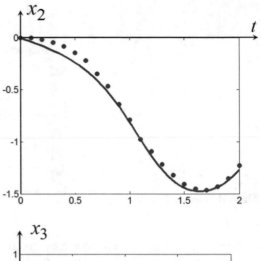

Fig. 5.22 The plot for $x_2(t)$

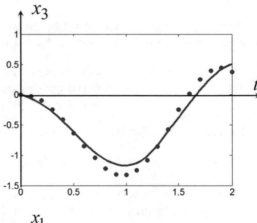

Fig. 5.23 The plot for $x_3(t)$

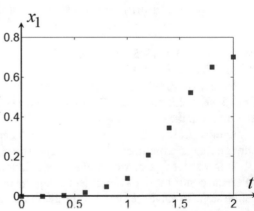

Fig. 5.24 The given trajectory for $x_1(t)$

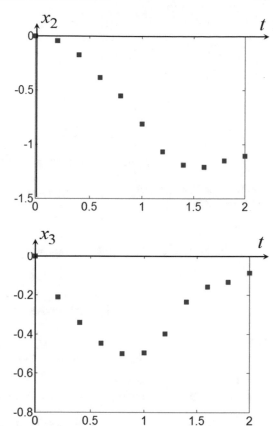

Fig. 5.25 The given trajectory for $x_2(t)$

Fig. 5.26 The given trajectory for $x_3(t)$

The deviation from a given trajectory was taken as a quality criterion for the synthesis problem

$$J = \sum_{k=1}^{M} \sqrt{\sum_{j=1}^{N} \sum_{i=1}^{3} (x_i(t_j, \mathbf{x}^{0,k}) - \tilde{x}_i(t_j))^2} \to \min_{\mathbf{u} \in U}, \qquad (5.69)$$

where M is a number of initial conditions, N is a number of points on the given trajectories, $M = 3$, $N = 10$.

In the experiments, the following initial conditions were set:

$$\mathbf{x}^{0,1} = [0\ 0\ 0]^T, \mathbf{x}^{0,2} = [-0.1\ -0.1\ -0.1]^T, \mathbf{x}^{0,3} = [0,1\ 0,1\ 0.1]^T.$$

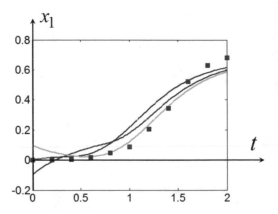

Fig. 5.27 Function $x_1(t)$ and the given trajectory

The basic solution was defined in the following form:

$$u_i = \begin{cases} u_i^-, & \text{if } \tilde{u}_i \leq u_i^-, \\ u_i^+, & \text{if } \tilde{u}_i \geq u_i^+, \\ \tilde{u}_i, & \text{otherwise} \end{cases}, \, i = 1,2,3, \tag{5.70}$$

where

$$\tilde{u}_i = c_1 x_1 + c_2 x_2 + c_3 x_3, \quad i = 1,2,3. \tag{5.71}$$

$c_i = 1, u_i^- = -2, u_i^+ = 2, i = 1,2,3.$

In the result, the following solution was obtained:

$$\begin{aligned}
\tilde{u}_1 &= (\cos(c_1)x_2 c_2 \cos(c_3))^2 - \sin(\sin(c_1) + \sin(x_1) + c_1), \\
\tilde{u}_2 &= (\cos(c_1)x_2 c_2 \cos(c_3))^2 - \sin(\sin(c_1) + \sin(x_1) + c_1) - \\
&\quad ((\cos(c_1)x_2 c_2 \cos(c_3))^2 - \sin(\sin(c_1) + \sin(x_1) + c_1))^3 + \\
&\quad c_1 + \cos(c_3 x_3) + \sin(\cos(x_3)\sin(\cos(c_1)c_2 x_2 \cos(c_3))), \\
\tilde{u}_3 &= \sqrt{c_1} + ((\cos(c_1)x_2 c_2 \cos(c_3))^2 - \sin(\sin(c_1) + \sin(x_1) + c_1))^3 + \\
&\quad \sin(\sqrt{c_1} + \cos(c_1) + c_2 x_2 \cos(c_3)),
\end{aligned} \tag{5.72}$$

where $c_1 = 0.01562, c_2 = 0.734375, c_3 = 2.484375$.

Figures 5.27, 5.28, and 5.29 show the results of obtained control system simulation with different initial values. Blue color is for initial values $\mathbf{x}^{0,1}$, red is for $\mathbf{x}^{0,2}$, and green is for $\mathbf{x}^{0,3}$.

As can be seen from Figs. 5.27, 5.28, and 5.29 the identified mathematical model allows synthesizing control system for multi-link robot that provides the movement over the given trajectory with different initial conditions.

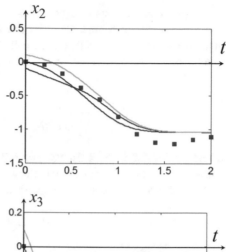

Fig. 5.28 Function $x_2(t)$ and the given trajectory

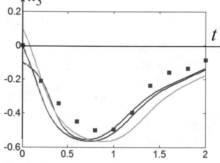

Fig. 5.29 Function $x_3(t)$ and the given trajectory

5.4 Synthesized Optimal Control Example

In this section, we are going to compare two approaches to practical implementation of the optimal control problem.

One way is to solve initially the optimal control problem and receive a control function in the form of time function. Then optimal trajectories are obtained basing on the received control function. To implement the solution, a stabilization control system is designed such that it can provide movement of the control object near optimal trajectory. The drawback of this approach is that it changes a mathematical model of control object after implementation of stabilization system and the initial optimal control is not more optimal for the modified object. We will further show this effect by introducing the noise into the object model.

Another approach of synthesized optimal control, described in Chap. 2 consists in initially creating a control system to provide stability to control object in some point of the state space. Then the optimal control problem is solved by finding the stabilization point positions. In the second approach, a mathematical model of control object is not changed after the solution of the optimal control problem.

Let us compare now two approaches on an example of the optimal control problem for two mobile robots. Consider the task of control of two mobile robots [2] that must swap without collisions with each other. The mathematical model of each robot is described by a system of $n = 3$ equations

$$
\begin{aligned}
\dot{x}_1^j &= 0.5(u_1^j + u_2^j)\cos(x_3^j) + B\xi, \\
\dot{x}_2^j &= 0.5(u_1^j + u_2^j)\sin(x_3^j) + B\xi, \\
\dot{x}_3^j &= 0.5(u_1^j - u_2^j) + B\xi,
\end{aligned} \tag{5.73}
$$

where j is the number of robots, $j = 1, 2$, B is a constant, determining a level of noise, ξ ξ is a random value from interval $(-1; 1)$.

There are constraints on components of the control vectors

$$
u_i^- \leq u_i^j \leq u_i^+, \tag{5.74}
$$

where u_i^-, u_i^+ are given values $i, j = 1, 2$.

The initial conditions are given

$$
x_i^j(0) = x_i^{0,j} + B_0\xi, \ i = 1, 2, 3, \ j = 1, 2, \tag{5.75}
$$

where B_0 is a noise level.

Terminal conditions are set

$$
x_i^j(t_f) = x_i^{f,j}, \ i = -1, 1, 2, 3, \ j = 1, 2. \tag{5.76}
$$

Each robot produces a dynamic phase constraint for the other robot. Collision avoidance condition is

$$
d^2 - (x_1^1 - x_1^2)^2 - (x_2^1 - x_2^2)^2 \leq 0, \tag{5.77}
$$

where d is a given minimal distance between centers of robots.

A quality criterion is given

$$
\begin{aligned}
J = t_f + a_1 \sum_{j=1}^{2} \sum_{1=1}^{3} \| x_i^j(t_f) - x_i^{f,j} \| + \\
a_2 \int_0^{t_f} \vartheta(d^2 - (x_1^1 - x_1^2)^2 - (x_2^1 - x_2^2)^2)dt \to \min,
\end{aligned} \tag{5.78}
$$

where a_1, a_2 are given weight coefficients,

$$
t_f = \max\{t_{f,1}, t_{f,2}\}, \tag{5.79}
$$

$$
t_{f,j} = \begin{cases} t, \text{ if } t < t^+ \text{ and } \delta_j(t) \leq \varepsilon_1, \\ t^+ \text{ otherwise}, \end{cases} \tag{5.80}
$$

where ε_1 is a small positive value, t^+ is a limit time of control process, $\vartheta(a)$ is a step Heaviside function

$$\vartheta(a) = \begin{cases} 1, \text{ if } a > 0 \\ 0 - \text{otherwise} \end{cases} ,$$

$$\delta_j(t) = \sqrt{\sum_{1=1}^{3} (x_i^j - x_i^{f,j})^2}. \tag{5.81}$$

The stated problem was solved by two different approaches: by the synthesized optimal control and by the direct approach using piece-wise linear approximation. In both approaches, control systems were constructed without disturbances at $B = 0$.

5.4.1 Synthesized Optimal Control

According to the approach, it is necessary to find a solution of the synthesized optimal control problem in the form of a set of coordinates of equilibrium points.

Firstly, the control system synthesis problem is solved. As far as the robots are similar, the problem of synthesis is solved for one robot. For solution of this problem, the symbolic regression method of variational Cartesian genetic programming was used.

In the result, the following control function was obtained:

$$u_i^j = \begin{cases} u_i^+ = u_i^+, \text{ if } u_i^+ \le \tilde{u}_i^j \\ u_i^- = u_i^-, \text{ if } \tilde{u}_i^j \le u_i^- \\ \tilde{u}_i^j, \text{ otherwise} \end{cases}, i = 1, 2, \ j = 1, 2, \tag{5.82}$$

where

$$\tilde{u}_1^j = A + B + \rho_\#(A), \ j = 1, 2, \tag{5.83}$$

$$\tilde{u}_1^j = B - A - \rho_\#(A), \ j = 1, 2, \tag{5.84}$$

$$A = c_1(\theta^* - \theta^j) + \sigma_\#((x^* - x^J)(y^* - y^J)), \tag{5.85}$$

$$B = 2(x^* - x^j) + \text{sgn}(x^* - x^J)c_2, \tag{5.86}$$

$$\rho_\#(\alpha) = \begin{cases} \text{sgn}(\alpha)B^+, \text{ if } |\alpha| > -\log(\delta^-) \\ \text{sgn}(\alpha)(\exp(|\alpha|) - 1) \end{cases}, \ \sigma_\#(\alpha) = \text{sgn}(\alpha)\sqrt{|\alpha|}, \tag{5.87}$$

$c_1 = 3.1094$, $c_2 = 3.6289$, $B^+ = 10^8$, $\delta^- = 10^{-8}$.

For solution of the synthesis problem, eight initial conditions were used and the quality criterion took into account the speed and the accuracy of terminal position achievement

$$\mathbf{x}^* = [x_1^* \ x_2^* \ x_3^*]^T. \tag{5.88}$$

In the result of the solution of control synthesis problem, a stable equilibrium point in the state space is appeared. Position of the equilibrium point depends on the terminal vector (5.88).

Secondly, the set of equilibrium points

$$X^* = \{x^{*,1,1}, \ldots, x^{*,1,4}, x^{*,2,1}, \ldots, x^{*,2,4}\} \tag{5.89}$$

was searched such that when switching from one point to another through given time interval Δ robots (5.73) moves from initial conditions to the terminal state with the optimal value of the quality criterion (5.78).

To search for the points the evolutionary algorithm of Grey wolf optimizer [7] was used. This algorithm differs from the most popular evolutionary algorithm of Particle Swarm Optimization [5] by not using such evolutionary parameters as a number of possible solutions in initial population and a number of generations as calculation parameters. GWO changes every possible solution basing on the information from three best current possible solutions. In the present work, GWO was slightly modified. Now a number of the best current possible solutions is a parameter of the algorithm. In calculations four or eight best current possible solutions are used. Parameters in the experiments were set as follows: dimension of the vector of parameters $p = 18$, restrictions on parameters $q^+_{1+3(j-1)} = 11$, $q^+_{2+3(j-1)} = 11$, $q^+_{3+3(j-1)} = \pi/2$, $q^-_{1+3(j-1)} = -1$, $q^-_{1+3(j-1)} = 1$, $q^-_{3+3(j-1)} = -\pi/2$, $j = 1, 2$. Initial conditions $x^{0,1} = [0\ 0\ 0]^T$, $x^{0,2} = [10\ 10\ 0]^T$, restrictions on control $u^+_i = 10$, $u^-_i = -10$, $i = 1, 2$, other parameters, control time limit $t^+ = 2.4$ s, interval for switching stabilization points $\Delta = 0.8$ s, $\varepsilon_1 = 0.01$, $d = 2$, $a_1 = 2.5$, $a_2 = 3.5$.

In result the following stabilization points were received:

$$\begin{aligned}
x^{*,1,1} &= [5.5\ 5.584\ 0.0744]^T, \\
x^{*,1,2} &= [9.1374\ 9.116\ 0.302]^T, \\
x^{*,1,3} &= [7.7585\ -0.4732\ 0.8357]^T, \\
x^{*,2,1} &= [0.4587\ 3.1368\ -0.4339]^T, \\
x^{*,2,2} &= [0.0726\ 0.5525\ 0.1014]^T, \\
x^{*,2,3} &= [0.3813\ -0.4348\ -0.7355]^T.
\end{aligned} \tag{5.90}$$

The value of functional (5.78) for found solution was $J = 2.1542$. In the Fig. 5.30 optimal trajectories for robots on the horizontal plane $\{x_1, x_2\}$ are presented. Black squares in the Fig. 5.30 are stabilization points.

5.4.2 Direct Solution of the Optimal Control Problem

Now the same problem is solved by traditional methodology, when the optimal control problem is solved firstly, and then the stabilization system is constructed in order to provide a steady movement of the object along the optimal trajectory.

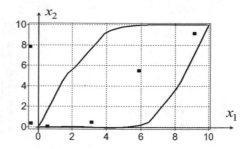

Fig. 5.30 Optimal trajectories
for synthesized control

Initially, for solution of the optimal control problem a piece-wise linear approximation of control was used. The control for both robots was calculated by the following equation:

$$u_i^j = \begin{cases} u_i^+, & \text{if } q(t,j,k,\Delta_0) > u_i^+ \\ u_i^-, & \text{if } q(t,j,k,\Delta_0) < u_i^- \\ q(t,j,k,\Delta_0), & \text{otherwise} \end{cases}, \quad j = 1,2, \ i = 1,2, \tag{5.91}$$

where

$$q(t,j,k,\Delta_0) = \left(q_{(i+2j-3)M+k+1} - q_{(i+2j-3)M+k}\right)\frac{t - \Delta_0(k-1)}{\Delta_0}, \tag{5.92}$$

$i = 1,2, \ j = 1,2, \ k = 1,\ldots,M-1,$

$$M = \left\lfloor \frac{t^+}{\Delta_0} \right\rfloor + 1.$$

During the solution of the optimal control problem, the following problem parameters were used: $t^+ = 2.4\,\text{s}$, $\Delta_0 = 0.24\,\text{s}$, $M = \lfloor t^+/\Delta_0 \rfloor + 1 = 11$. Restrictions on parameter values were $q^+ = 20$, $q^- = -20$. Optimal values of parameters were again searched by the GWO algorithm.

As a result, the following values of parameters were received:

$q_1 = 19.0344$, $q_2 = 18.7579$, $q_3 = 14.8563$, $q_4 = 16.252$, $q_5 = 13.9701$, $q_6 = 14.2058$, $q_7 = 12.1852$, $q_8 = 0.5533$, $q_9 = -0.6065$, $q_{10} = -1.0168$, $q_{11} = -0.758$, $q_{12} = -2.052$, $q_{13} = 19.7234$, $q_{14} = 11.9609$, $q_{15} = 13.3309$, $q_{16} = -2.0296$, $q_{17} = 17.6389$, $q_{18} = 10.0327$, $q_{19} = 17.9334$, $q_{20} = 2.5053$, $q_{21} = 1.174$, $q_{22} = 0.2601$, $q_{23} = 8.9844$, $q_{24} = -17.0169$, $q_{25} = -17.7283$, $q_{26} = -15.8004$, $q_{27} = -17.1493$, $q_{28} = -14.0442$, $q_{29} = -19.1540$, $q_{30} = 0.0241$, $q_{31} = 2.0056$, $q_{32} = 0.3253$, $q_{33} = -0.2754$, $q_{34} = -17.4827$, $q_{35} = -12.7697$, $q_{36} = -10.3852$, $q_{37} = -15.7939$, $q_{38} = -11.4048$, $q_{39} = -18.0859$ $q_{40} = -2.5518$, $q_{41} = -0.1689$, $q_{42} = 0.1689$, $q_{43} = 4.8284$ $q_{44} = -2.6376$.

A value of functional was obtained $J = 1.9013$.

Figure 5.31 shows optimal trajectories received by the direct approach.

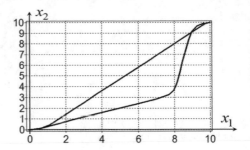

Fig. 5.31 Optimal trajectories received by the direct approach

Thus, the solution of the optimal control problem (5.91) is a function of time. Obviously, this solution could not be realized on board of robots. So next, for implementation of the received solution it is necessary to construct a stabilization system.

Generally speaking, the construction of the stabilization system is very dependent on the specific object.

In the present example, the stabilization system for moving robots on optimal trajectories is constructed according to [8].

For stabilization, the following control is used:

$$
\begin{aligned}
u_1^j &= 0.5(u_1^{j*} + u_2^{j*})(\cos(e_3^j) + k_2^j e_2^j) + k_1^j e_1^j + \\
&\quad 0.5(u_1^{j*} - u_2^{j*}) + k_3^j \sin(e_3^j), \\
u_2^j &= 0.5(u_1^{j*} + u_2^{j*})(\cos(e_3^j) - k_2^j e_2^j) + k_1^j e_1^j - \\
&\quad 0.5(u_1^{j*} - u_2^{j*}) - k_3^j \sin(e_3^j),
\end{aligned}
\tag{5.93}
$$

where upper index j is the robot number, u_1^j and u_2^j are the optimal control for robot j, e_1^j, e_2^j, e_3^j are errors between state of robot j and its optimal state,

$$
\begin{bmatrix} e_1^j \\ e_2^j \\ e_3^j \end{bmatrix} =
\begin{bmatrix} k_2^j \cos(x_3^j) & k_2^j \sin(x_3^j) & 0 \\ -k_2^j \sin(x_3^j) & k_2^j \cos(x_3^j) & \alpha^j \\ 0 & 0 & 1 \end{bmatrix}
\begin{bmatrix} x_1^j * - x_1^j \\ x_2^j * - x_2^j \\ x_3^j * - x_3^j \end{bmatrix},
\tag{5.94}
$$

$x_1^{j*}, x_2^{j*}, x_3^{j*}$ are components of optimal state vector for robot j, they are determined from the etalon model

$$
\begin{aligned}
\dot{x}_1^{j*} &= 0.5(u_1^{j*} + u_2^{j*})\cos(x_3^{j*}), \\
\dot{x}_2^{j*} &= 0.5(u_1^{j*} + u_2^{j*})\sin(x_3^{j*}), \\
\dot{x}_3^{j*} &= 0.5(u_1^{j*} - u_2^{j*}),
\end{aligned}
\tag{5.95}
$$

$k_1^j, k_2^j, k_3^j, \alpha^j$ are constant coefficients of controller, $j = 1, 2$.

To search for optimal coefficients $k_1^j, k_2^j, k_3^j, \alpha^j$ the same GWO was used. Values of parameters are determined on the criterion (5.78).

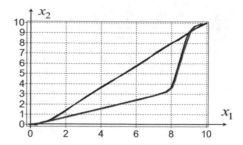

Fig. 5.32 The optimal trajectories with and without stabilization system

In the result, the following values were received $k_1^1 = 9.9554$, $k_2^1 = 5.9981$, $k_3^1 = 7.1414$, $\alpha^1 = 3.3537$, $k_1^2 = 3.0468$, $k_2^2 = 6.3653$, $k_3^2 = 8.8845$, $\alpha^2 = 0.6517$. The resulting value of the quality criterion (5.78) was improved a little $J = 1.8968$.

The optimal trajectories for the object with the stabilization system and optimal trajectories without the stabilization system are presented in Fig. 5.32.

In the Fig. 5.32 red lines are optimal trajectories for the etalon model, and black ones are optimal trajectories for the object with the stabilization system. As can be seen, trajectories with the stabilization system almost coincide with the etalon model completely.

5.4.3 Experimental Analysis of Sensitivity to Perturbations

To compare two approaches, the synthesized optimal control and direct optimal control with realized stabilization system, the perturbations were entered into the mathematical model and in the initial conditions.

There were ten experiments for each perturbation level. Values of the functional with different perturbations are presented in Tables 5.4, 5.5, 5.6, and 5.7.

Table 5.4 Direct approach with stabilization in the presence of model perturbations

No	$B = 0.5$	$B = 1$	$B = 2$	$B = 5$
1	3.4768	5.0309	4.1677	6.6225
2	3.0823	3.2038	7.3429	4.9321
3	3.0805	4.9694	4.1394	6.9220
4	3.1019	3.1120	8.2476	7.2693
5	5.1297	4.8077	3.1618	9.7189
6	3.0917	2.6837	4.9604	8.2018
7	3.2268	4.9566	3.2740	8.7121
8	3.7074	5.3586	3.0518	5.1837
9	5.4770	3.3332	7.2745	4.6449
10	3.6222	5.1715	8.6932	8.9466
Average	3.6969	4.2627	5.4313	7.1234
St.dev.	0.8803	1.0380	2.2255	1.7793

Table 5.5 Synthesized optimal control in the presence of model perturbations

No	$B = 0.5$	$B = 1$	$B = 2$	$B = 5$
1	2.5524	2.5413	2.8345	2.6035
2	2.5019	2.6449	2.7889	3.5243
3	2.5056	2.4866	2.6103	3.2268
4	2.4952	2.6251	2.7708	3.1393
5	2.4856	2.5473	2.6309	3.6660
6	2.4965	2.2293	2.7764	3.4443
7	2.2000	2.6381	2.7795	3.4076
8	2.4517	2.6211	2.8168	2.9567
9	2.4861	2.7025	2.5121	3.3035
10	2.4878	2.5452	2.7744	3.4872
Average	2.4663	2.55814	2.7178	3.3506
St.dev.	0.0968	0.13209	0.158	0.2172

Table 5.6 Direct approach with stabilization in the presence of perturbations in initial conditions

No	$B_0 = 0.1$	$B_0 = 0.5$	$B_0 = 1$	$B_0 = 1.5$	$B_0 = 2$
1	3.0550	3.0230	2.8399	2.6057	3.1508
2	2.5150	2.5412	2.5558	2.8344	7.7675
3	2.5249	2.9201	2.6229	31.2255	2.6863
4	3.0181	2.5976	2.6259	2.9744	2.8958
5	2.8731	2.5149	2.5205	2.5147	2.9021
6	2.5164	2.5142	2.9247	2.6986	2.5765
7	2.5146	2.7903	2.5227	2.5148	49.2804
8	2.5154	2.6121	2.5151	2.8189	9.5292
9	2.5215	2.5358	2.8036	2.5146	9.9916
10	2.6112	2.6212	2.5464	2.7979	9.3691
Average	2.6665	2.667	2.6478	5.5501	9.3691
St.dev.	0.2243	0.1812	0.1517	9.0228	13.6044

As can be seen form the Tables 5.4, 5.5, 5.6, and 5.7, for big perturbations, $B_0 = 1.5$ or $B_0 = 2$, there appeared some enormous values of the functional for the direct approach of optimal control with stabilization system. For example, in the experiment 3 with $B_0 = 1.5$ the functional was 31.2255, or in the experiment 7 with $B_0 = 2$ the functional was 49.2804. These were not mistakes. These cases always appeared at big perturbations. This means that the stabilization system is corrupted and could not provide the movement of robots to the terminal points near the optimal trajectories. For the synthesized control at the same perturbations, such big values of functional did not appear.

Also, when discussing the obtained results, we would note one more weakness of the direct approach. In the Fig. 5.33 trajectories of robots with perturba-

Table 5.7 Synthesized optimal control in the presence of perturbations in initial conditions

No	$B_0 = 0.1$	$B_0 = 0.5$	$B_0 = 1$	$B_0 = 1.5$	$B_0 = 2$
1	2.4601	2.4512	2.535	2.633	2.526
2	2.4816	2.4929	2.7775	2.5091	4.3804
3	2.4635	2.4757	2.5011	7.9801	12.845
4	2.4565	2.4868	2.5107	4.7016	2.5051
5	2.4858	2.4846	2.5724	2.5074	7.214
6	2.4643	2.5051	2.501	2.525	2.4933
7	2.4605	2.501	2.4853	2.9441	15.7325
8	2.455	2.5422	2.7048	6.5336	16.6373
9	2.4804	2.6022	2.5551	3.0706	2.5289
10	2.4854	2.4644	2.4625	2.4656	3.7654
Average	2.4693	2.5006	2.5606	3.78701	7.0628
St.dev.	0.0125	0.0434	0.1019	1.9756	5.7846

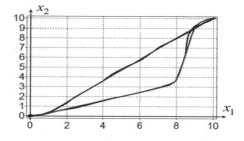

Fig. 5.33 Optimal trajectories received by the direct approach

tions $B = 2$ are presented. Optimal trajectories signed red color almost coincides with trajectories of objects' movement, but value of the functional (5.78) for movement with perturbations is equal to 3.274 instead of 1.8968 for optimal trajectories. This confirms the proposition that movement near optimal trajectory is not optimal movement in terms of functional value, it gives not optimal value of functional.

5.5 Machine Learning in Synergetic Control

In this example, we will consider the possibility of using machine learning methods based on symbolic regression to synthesize synergetic control. The application of the ideas of Synergetics [9] to the solution of nonlinear control problems was proposed in [10, 11] and is considered as synergetic control which is characterized by the presence in the state space of such manifolds that must have special properties of attraction or repelling. The main idea of the approach is that when a feedback control function is synthesized and inserted into the model, all solutions of the resulting system of differential equations have special properties specified by the developer. Such properties that the solutions must satisfy are determined by the presence of manifolds in the state space. In particular, terminal manifolds must have the properties of an attractor, and phase constraints on the contrary must indicate regions of the state space through which the solutions of the system do not pass or have the repeller properties.

Consider a problem where the terminal manifold should have an attractor property.

The problem statement of the control synthesis with attracting terminal manifold is as follows.

The mathematical model of control object is given

$$\dot{\mathbf{x}} = \mathbf{f}(\mathbf{x}, \mathbf{u}), \tag{5.96}$$

where $\mathbf{x} \in \mathbb{R}^n$, $\mathbf{u} \in U \subseteq \mathbb{R}^m$, U is a compact set, $m \leq n$.

The domain of initial conditions is given

$$X_0 \subseteq \mathbb{R}^n. \tag{5.97}$$

The terminal manifold is given

$$\mathbf{g}(\mathbf{x}(t)) = 0, \tag{5.98}$$

where $t \geq t_f$, t_f is a time of reaching the manifold (5.98),

$$\mathbf{g}(\mathbf{x}) = [g_1(\mathbf{x}) \ldots g_r(\mathbf{x})]^T, \ r \leq n. \tag{5.99}$$

The quality criterion is given

$$J = \int_0^{t_f} f_0(\mathbf{x}(t), \mathbf{u}(t)) dt \to \min_{\mathbf{u} \in U}. \tag{5.100}$$

It is necessary to find a control function in the form

$$\mathbf{u} = \mathbf{h}(\mathbf{x}). \tag{5.101}$$

When this control function is substituted into the right part of the system of differential equations (5.96), then any solution of the system

$$\dot{\mathbf{x}} = \mathbf{f}(\mathbf{x}, \mathbf{h}(\mathbf{x})) \tag{5.102}$$

from the initial conditions from the given region (5.97) will reach the terminal manifold (5.99) and remain on it after reaching and the value of the quality criterion (5.102) will be optimal.

For numerical solution of the problem, it is necessary to reformulate the statement in order to provide computer checking of the given conditions.

Replace the initial condition region with a set of initial condition points

$$\tilde{X}_0 = \{\mathbf{x}^{0,1}, \ldots, \mathbf{x}^{0,K}\}. \tag{5.103}$$

In order to check the attraction property condition of the terminal manifold (5.98), a multi-point criterion is introduced. A certain number M of points on the terminal manifold is given. A position of points is not determined, but only the number of points. Any partial solution of the system (5.102) from the given set of initial conditions (5.103) must pass a given number of points on the manifold. Distance between any pair of points on the manifold must be not less than the given value δ, at that when moving from one point to the next, the partial solution must not deviate from the manifold by more than the given value ε. Multi-point criterion for checking the conditions of attraction of the terminal manifold is included into the quality functional.

$$J_1 = \sum_{i=1}^{K} \left(\int_0^{t^+} f_0(\mathbf{x}(t, \mathbf{x}^{0,i}), \mathbf{h}(\mathbf{x}(t, \mathbf{x}^{0,i}))) dt + \right.$$

$$\left. p_1(M - \min\{M, L\}) + p_2 \sqrt{\sum_{j=1}^{r} g_j(\mathbf{x}(t^+, \mathbf{x}^{0,i}))} \right), \tag{5.104}$$

where $\mathbf{x}(t, \mathbf{x}^{0,i})$ is a partial solution of the system (5.102) from initial conditions $\mathbf{x}^{0,i}$, $i \in \{1, \ldots, K\}$, t^+ is a given time, p_1, p_2 are given weight coefficients, L is a cardinal number of the set of points on the terminal manifold $T(\mathbf{x}(t, \mathbf{x}^{0,i}))$

$$T(\mathbf{x}(t, \mathbf{x}^{0,i})) = \{\mathbf{x}(t_1, \mathbf{x}^{0,i}), \ldots, \mathbf{x}(t_L, \mathbf{x}^{0,i})\}, \tag{5.105}$$

$t_1 < \ldots < t_L$,

$$\|\mathbf{x}(t_k, \mathbf{x}^{0,i}) - \mathbf{x}(t_s, \mathbf{x}^{0,i})\| \geq \delta, \forall k, s \in \{1, \ldots, L\}, \ k \neq s, \tag{5.106}$$

$$\sqrt{\sum_{j=1}^{r} g_j(\mathbf{x}(t^+, \mathbf{x}^{0,i}))} \leq \varepsilon, \ t_1 \leq t \leq t_L. \tag{5.107}$$

Consider an Example of the Synergistic Control Synthesis

The mathematical model of control object has the following form:

$$\begin{aligned} \dot{x}_1 &= x_2 + u_1, \\ \dot{x}_2 &= -x_2 - x_1 - x_1^3 + u_2. \end{aligned} \tag{5.108}$$

Restrictions on control are given

$$-1 \le u_i \le 1, \ i = 1,2. \tag{5.109}$$

The terminal manifold is described by the following equation:

$$x_2 = \frac{x_{2,2} - x_{2,1}}{x_{1,2} - x_{1,1}} x_1, \ \text{if} \ x_{1,1} \ne x_{1,2}, \tag{5.110}$$

$$x_1 = 0, \ \text{if} \ x_{1,1} = x_{1,2}, \tag{5.111}$$

$$x_i^- = \min\{x_{i,1}, x_{i,2}\} \le x_i \le \max\{x_{i,1}, x_{i,2}\} = x_i^+, \ i = 1,2. \tag{5.112}$$

To obtain the attractor property, the quality criterion should define a distance to the terminal manifold. Therefore, the quality criterion has the following form:

$$J_2 = \sum_{i=1}^{K} \left(p_1(M - \min\{M, L\}) + p_2 \Delta_f(t^+, \mathbf{x}^{0,i}) + \int_0^{t^+} \Delta_f(t, \mathbf{x}^{0,i}) dt \right), \tag{5.113}$$

where $\Delta_f(t, \mathbf{x}^{0,i})$ is a distance from partial solution $\mathbf{x}(t, \mathbf{x}^{0,i})$ to the terminal manifold in the moment t

$$\Delta_f(t, \mathbf{x}^{0,i}) = \begin{cases} \min\{\Delta^-(t, \mathbf{x}^{0,i}), \Delta^+(t, \mathbf{x}^{0,i})\}, \\ \quad \text{if} \ (\beta^-(t) < 0) \vee (\beta^+(t) < 0); \\ 2S(t)/l_f - \text{otherwise,} \end{cases}$$

$$\Delta^-(t, \mathbf{x}^{0,i}) = \sqrt{(x_1(t, \mathbf{x}^{0,i}) - x_1^-)^2 + (x_2(t, \mathbf{x}^{0,i}) - x_2^-)^2},$$

$$\Delta^+(t, \mathbf{x}^{0,i}) = \sqrt{(x_1(t, \mathbf{x}^{0,i}) - x_1^+)^2 + (x_2(t, \mathbf{x}^{0,i}) - x_2^+)^2},$$

$$\beta^-(t) = \sum_{i=1}^{2} (x_i^+ - x_i^-)(x_i(t, \mathbf{x}^{0,i}) - x_i^-),$$

$$\beta^+(t) = \sum_{i=1}^{2} (x_i^- - x_i^+)(x_i(t, \mathbf{x}^{0,i}) - x_i^+),$$

$$l_f = \sqrt{(x_1^+ - x_1^-)^2 + (x_2^+ - x_2^-)^2},$$

$$S(t) = \sqrt{p(t)(p(t) - \Delta^+(t, \mathbf{x}^{0,i}))(p(t) - \Delta^-(t, \mathbf{x}^{0,i}))(p(t) - l_f)},$$

$$p(t) = 0.5(\Delta^-(t, \mathbf{x}^{0,i}) + \Delta^+(t, \mathbf{x}^{0,i}) + l_f).$$

In the experiments the set of initial conditions included $K = 16$ points:

$$\tilde{X}_0 = \{x^{0,1} = [-1.1 \ -1.1]^T, x^{0,2} = [-1.1 \ -0.4]^T,$$
$$x^{0,3} = [-1.1 \ 0.3]^T, x^{0,4} = [-1.1 \ 1]^T, x^{0,5} = [-0.4 \ -1.1]^T,$$
$$x^{0,6} = [-0.4 \ -0.4]^T, x^{0,7} = [-0.4 \ 0.3]^T, x^{0,8} = [-0.4 \ 1]^T,$$
$$x^{0,9} = [0.3 \ -1.1]^T, x^{0,10} = [0.3 \ -0.4]^T, x^{0,11} = [0.3 \ 0.3]^T,$$
$$x^{0,12} = [0.3 \ 1]^T, x^{0,13} = [1 \ -1.1]^T, x^{0,14} = [1 \ -0.4]^T,$$
$$x^{0,15} = [1 \ 0.3]^T, x^{0,16} = [1 \ 1]^T\}. \tag{5.114}$$

Other parameters of the task had the following values: $p_1 = 0.01$, $p_2 = 1$, $M = 8$, $\varepsilon = 0.005$, $\delta = 0.005$, $t^+ = 4$ s.

For solution of this problem, the multi-layer network operator was used.

The basic solution had the following form:

$$u_i = \begin{cases} -1, \text{ if } \tilde{u}_i \leq -1 \\ 1, \text{ if } \tilde{u}_i \geq 1 \\ \tilde{u}_i, \text{ otherwise} \end{cases}, \tag{5.115}$$

where

$$\tilde{u}_1 = -q_3(q_1x_1 + q_2x_2) + q_4 - q_1x_1 - q_2x_2,$$
$$\tilde{u}_2 = -q_3(q_1x_1 + q_2x_2) + q_4 - q_1x_1 - q_2x_2, \tag{5.116}$$

$q_i = 1$, $i = 1, 2, 3, 4$.

The basic solution (5.115),(5.116) was coded by the network operator of $N = 2$ layers of the dimension 16×16:

$$\Psi^1 = \begin{bmatrix} 0&0&0&0&1&0&0&0&0&0&0&0&0&0&0&0 \\ 0&0&0&0&1&0&0&0&0&0&0&0&0&0&0&0 \\ 0&0&0&0&0&1&0&0&0&0&0&0&0&0&0&0 \\ 0&0&0&0&0&1&0&0&0&0&0&0&0&0&0&0 \\ 0&0&0&0&2&0&3&0&0&0&0&0&0&0&0&0 \\ 0&0&0&0&0&2&3&0&0&0&0&0&0&0&0&0 \\ 0&0&0&0&0&0&1&1&0&0&0&0&0&0&0&0 \\ 0&0&0&0&0&0&1&1&0&0&0&0&0&0&0&0 \\ 0&0&0&0&0&0&0&1&1&0&0&0&0&0&0&0 \\ 0&0&0&0&0&0&0&0&1&1&0&0&0&0&0&0 \\ 0&0&0&0&0&0&0&0&0&1&1&0&0&0&0&0 \\ 0&0&0&0&0&0&0&0&0&0&1&1&0&0&0&0 \\ 0&0&0&0&0&0&0&0&0&0&0&1&1&0&0&0 \\ 0&0&0&0&0&0&0&0&0&0&0&0&1&1&0&0 \\ 0&0&0&0&0&0&0&0&0&0&0&0&0&1&1&0 \\ 0&0&0&0&0&0&0&0&0&0&0&0&0&0&1&1 \\ 0&0&0&0&0&0&0&0&0&0&0&0&0&0&0&1 \end{bmatrix}, \tag{5.117}$$

$$\Psi^2 = \begin{bmatrix}
0 & 0 & 0 & 0 & 1 & 0 & 0 & 0 & 0 & 0 & 0 & 0 & 0 & 0 & 0 \\
0 & 0 & 0 & 0 & 1 & 0 & 0 & 0 & 0 & 0 & 0 & 0 & 0 & 0 & 0 \\
0 & 0 & 0 & 0 & 0 & 1 & 0 & 0 & 0 & 0 & 0 & 0 & 0 & 0 & 0 \\
0 & 0 & 0 & 0 & 0 & 1 & 0 & 0 & 0 & 0 & 0 & 0 & 0 & 0 & 0 \\
0 & 0 & 0 & 2 & 0 & 1 & 0 & 0 & 0 & 0 & 0 & 0 & 0 & 0 & 0 \\
0 & 0 & 0 & 0 & 0 & 1 & 1 & 0 & 0 & 0 & 0 & 0 & 0 & 0 & 0 \\
0 & 0 & 0 & 0 & 0 & 0 & 1 & 1 & 0 & 0 & 0 & 0 & 0 & 0 & 0 \\
0 & 0 & 0 & 0 & 0 & 0 & 0 & 1 & 1 & 0 & 0 & 0 & 0 & 0 & 0 \\
0 & 0 & 0 & 0 & 0 & 0 & 0 & 0 & 1 & 1 & 0 & 0 & 0 & 0 & 0 \\
0 & 0 & 0 & 0 & 0 & 0 & 0 & 0 & 0 & 1 & 1 & 0 & 0 & 0 & 0 \\
0 & 0 & 0 & 0 & 0 & 0 & 0 & 0 & 0 & 0 & 1 & 1 & 0 & 0 & 0 \\
0 & 0 & 0 & 0 & 0 & 0 & 0 & 0 & 0 & 0 & 0 & 1 & 1 & 0 & 0 \\
0 & 0 & 0 & 0 & 0 & 0 & 0 & 0 & 0 & 0 & 0 & 0 & 1 & 1 & 0 \\
0 & 0 & 0 & 0 & 0 & 0 & 0 & 0 & 0 & 0 & 0 & 0 & 0 & 1 & 1 \\
0 & 0 & 0 & 0 & 0 & 0 & 0 & 0 & 0 & 0 & 0 & 0 & 0 & 0 & 1
\end{bmatrix}. \tag{5.118}$$

Input and output vectors of this network operator were

$$\mathbf{r}^1 = [0\ 0\quad 0\ 2\quad 0\ 1\quad 0\ 3]^T, \tag{5.119}$$

$$\mathbf{r}^2 = [1\ 14\quad 0\ 4\quad 1\ 15\quad 0\ 5]^T, \tag{5.120}$$

$$\mathbf{d}^1 = [2\ 14]^T, \tag{5.121}$$

$$\mathbf{d}^2 = [2\ 15]^T. \tag{5.122}$$

Consider the terminal manifold in the form of a segment in various positions. Let firstly the terminal manifold be described by two points

$$(x_{1,1} = -0.5, x_{2,1} = -0.5), (x_{1,2} = 0.5, x_{2,2} = 0.5). \tag{5.123}$$

In the result of computations, the following solution was received:

$$\begin{aligned}
\tilde{u}_1 &= q_3(x_2 - 2q_1x_1 - q_2x_2 + \text{sgn}(x_2 - q_1x_1)\sqrt{|x_2 - q_1x_1|}) + \\
&\quad q_4(x_2 - 2q_1x_1 - q_2x_2 + \text{sgn}(x_2 - q_1x_1)\sqrt{|x_2 - q_1x_1|} + \\
&\quad \text{sgn}(x_2 - q_1x_1 - q_2x_2) + (q_2x_2)^3), \\
\tilde{u}_2 &= \tilde{u}_1 - \tilde{u}_1^3,
\end{aligned} \tag{5.124}$$

where $q_1 = 0.989258$, $q_2 = 0.024414$, $q_3 = 3.755859$, $q_4 = 0.106445$.

Figures 5.34 and 5.35 show the results of simulation of the control object (5.102) with the control function (5.124).

As seen from the results of simulation, the terminal manifold possess an attractor property: the object reaches the manifold and moves along it.

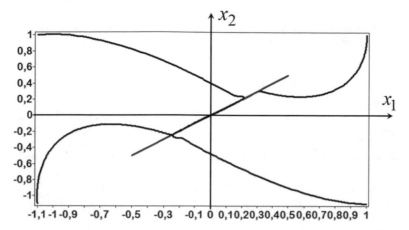

Fig. 5.34 Partial solutions of the system (5.102) for four initial conditions from the set $\tilde{X}_0$

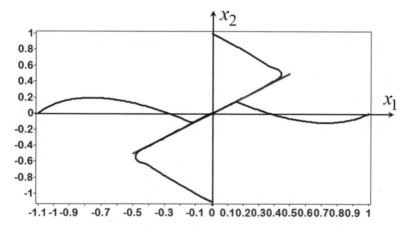

Fig. 5.35 Partial solutions of the system (5.102) for four initial conditions not from the set $\tilde{X}_0$

Consider now the terminal manifold based on two other points

$$(x_{1,1} = -0.5, x_{2,1} = 0,5), (x_{1,2} = 0.5, x_{2,2} = -0.5). \qquad (5.125)$$

For the terminal manifold (5.125), the following control function was received:

$$
\begin{aligned}
\tilde{u}_1 &= \arctan(+q_4(\arctan(-q_1 x_1 - q_2 \sin(x_2)) + x_2^3) + \\
&\quad -q_3(q_1 x_1 + q_2 \sin(x_2)) + (\arctan(-q_1 x_1 - q_2 \sin(x_2)) + x_2^3)^3), \\
\tilde{u}_2 &= \tilde{u}_1 + \vartheta(\tilde{u}_1) + \sin(-q_3(q_1 x_1 + q_2 \sin(x_2)) + \\
&\quad q_4(\arctan(-q_1 x_1 - q_2 \sin(x_2)) + x_2^3)),
\end{aligned}
\qquad (5.126)
$$

where $q_1 = 2.578125$, $q_2 = 2.665039$, $q_3 = 3.252930$, $q_4 = 1.182617$.
The results of simulation are shown in Figs. 5.36 and 5.37.

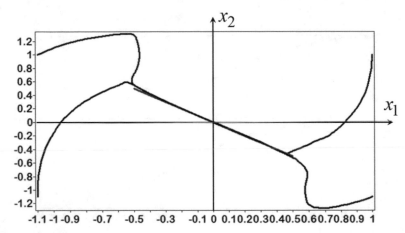

Fig. 5.36 Partial solutions of the system (5.102) with control (5.126) for four initial conditions from the set $\tilde{X}_0$

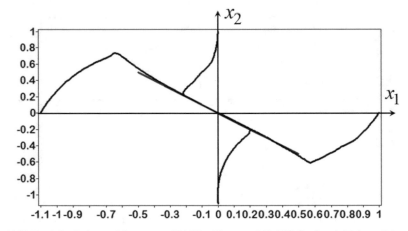

Fig. 5.37 Partial solutions of the system (5.102) with control (5.126) for four initial conditions that are not from the set $\tilde{X}_0$ of initial conditions

Consider one more terminal manifold based on two other points

$$(x_{1,1} = -0.5, x_{2,1} = 0), (x_{1,2} = 0.5, x_{2,2} = 0). \tag{5.127}$$

For the terminal manifold (5.127), the following control function was received:

$$\tilde{u}_1 = q_3 B + q_4 (B + \vartheta(x_2) + \arctan(q_1 x_1)) + \arctan(q_4),$$

$$\tilde{u}_2 = \sin(q_3 B + q_4 (B + \vartheta(x_2) + \arctan(q_1 x_1)) + \arctan(q_4)) \\ - B - \vartheta(x_2) - \arctan(q_1 x_1), \tag{5.128}$$

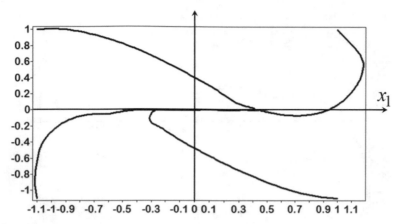

Fig. 5.38 Partial solutions of the system (5.102) with the control function (5.128) for four initial conditions from the set $\tilde{X}_0$

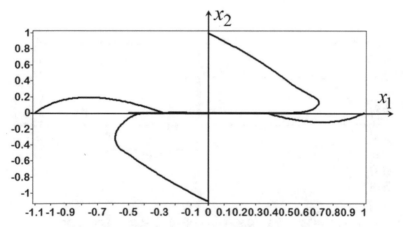

Fig. 5.39 Partial solutions of the system (5.102) with the control function (5.128) for four initial conditions not from the set $\tilde{X}_0$

where

$$B = A + A^3 + \arctan(-x_2 - q_1 x_1 + \sin(x_2) + \sqrt[3]{x_2}),$$
$$A = -q_1 x_1 + \sin(x_2) + \sqrt[3]{x_2} + \mathrm{sgn}(x_2)\sqrt{|x_2|},$$

$q_1 = 2.458008$, $q_2 = 0.795898$, $q_3 = 0.337891$, $q_4 = 0.283203$.

Figures 5.38 and 5.39 show the results of simulation of the system (5.102) with obtained control (5.128) from different initial conditions.

At last consider the terminal manifold

$$(x_{1,1} = 0, x_{2,1} = -0.5), (x_{1,2} = 0, x_{2,2} = 0.5). \tag{5.129}$$

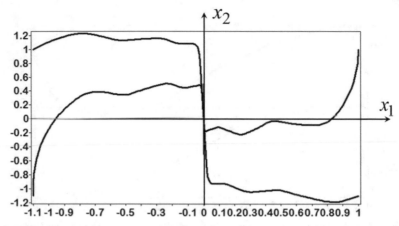

Fig. 5.40 Partial solutions of the system (5.102) with the control function (5.130) for four initial conditions from the set $\tilde{X}_0$

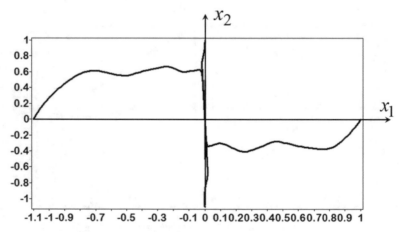

Fig. 5.41 Partial solutions of the system (5.102) with the control function (5.130) for four initial conditions not from the set $\tilde{X}_0$

For the terminal manifold (5.125), the following control function was received by the multi-layer network operator method:

$$\tilde{u}_1 = A + 2\tanh(A),$$

$$\tilde{u}_2 = A + 2\tanh(A) + 2\sin(A + 2\tanh(A)) + \vartheta(A + 2\tanh(A)),$$

(5.130)

where

$$A = -q_3(q_1 x_1 + q_2 x_2) + q_4 \sin(-q_1 x_1 - q_2 x_2),$$

$q_1 = 3.939453$, $q_2 = 0.204102$, $q_3 = 3.364258$, $q_4 = 1.684570$.

The results of simulation of the system (5.102) with obtained control (5.130) from different initial conditions are presented in Figs. 5.40 and 5.41.

References

1. Pontryagin, L.S., Boltyanskii, V.G., Gamkrelidze, R.V., Mishchenko, E.F.: The Mathematical Theory of Optimal Process. Gordon and Breach Science Publishers, New York/London/Paris/Montreux/Tokyo (1985)
2. Šuster, P., Jadlovská, A.: Tracking trajectory of the mobile robot Khepera II using approaches of artificial intelligence. Acta Electrotechnica et Informatica 11, 38–43 (2011)
3. Guryanov, A.E.: Quadrocopter control modeling. Engineering Bulletin. Bauman Moscow State Technical University, 2014, C. 522–534. (in Russian)
4. Liu, X.L., Duan, G.R., Teo, K.L.: Optimal soft landing control for moon lander. Automatica 44, 1097–1103 (2008)
5. Kennedy, J., Eberhart, R.: Particle swarm optimization. In: Proceedings of ICNN'95 – International Conference on Neural Networks, vol. 4, pp. 1942–1948 (1995)
6. Dang, T.P., Diveev, A.I., Kazaryan, D.E., Sofronova, E.A.: Identification control synthesis by the network operator method. In: Proceedings 2015 IEEE 10th Conference on Industrial Electronics and Applications (ICIEA), pp. 1559–1564 (2015)
7. Mirjalili, S., Mirjalili, S.M., Lewis, A.: Grey wolf optimizer. Adv. Eng. Softw. 69, 46–61 (2014)
8. Avendaño-Jurarez, J.L., Hernràndez-Guzmràn, V.M., Silva-Ortigoza, R.: Velocity and current inner loops in a wheeled mobile robot. Adv. Robot. 24(8–9), 1385–1404 (2010)
9. Haken, H.: Synergetics. Introduction and Advanced Topics. Springer, Berlin (2004)
10. Kolesnikov, A.A.: Introduction of synergetic control. In: 2014 American Control Conference (ACC), Portland (2014)
11. Kolesnikov, A.A., Kolesnikov, A.A.: Synergetic control theory and vibro-mechanics: a conceptual relation. Mekhatronika, Avtomatizatsiya, Upravlenie 16(5), 291–299 (2015) (In Russ.) https://doi.org/10.17587/mau.16.291-299

Printed in the United States
by Baker & Taylor Publisher Services